Nothing I See Means Anything

Quantum Questions, Quantum Answers

David Parrish, MD

Sentient Publications

First Sentient Publications edition 2006

A paperback original

Printed in the United States of America

Cover design by Kim Johansen, Black Dog Design
Book design by Nicholas Cummings

Library of Congress Cataloging-in-Publication Data

Parrish, David, MD.
Nothing I see means anything : quantum questions, quantum answers / David Parrish.— 1st Sentient Publications ed.
p. cm.
ISBN 1-59181-039-6
1. Consciousness. 2. Quantum theory. 3. Mysticism. I. Title.

BF311.P31368 2005
150—dc22

2005029685

10 9 8 7 6 5 4 3 2 1

SENTIENT PUBLICATIONS
A Limited Liability Company
1113 Spruce Street
Boulder, CO 80302
www.sentientpublications.com

In Honor of Michel Eyquem de Montaigne
(1533 – 1592)

"We are born to inquire after Truth"...it belongs to a greater power to possess it. It is not as Democritus said, hid in the bottom of the deeps, but rather elevated to an infinite height in the divine knowledge!

—Michel Eyquem de Montaigne

Nothing I See Means Anything

The reason this is so is that I see nothing, and nothing has no meaning. It is necessary that I recognize this that I may learn to see. What I think I see now is taking the place of vision. I must let it go by realizing it has no meaning, so that vision may take its place.

—*A Course in Miracles*

Contents

Foreword

When I first picked up this book, the title informed me that "nothing I see means anything." That obviously included the book. Why, then, read it if it had no meaning? But like so many of the penetrating insights that David Parrish has woven into these pages, I discovered that it is necessary to consider this one as thoughtfully as he presents it.

There is a profound difference between saying that no *thing* has meaning and that *seeing* has no meaning. As is shown through Dr. Parrish's brilliant examination of the phenomenon of projection, the way we see the things—the people, the circumstances—around us, holds the key to all meaning. Our take on the world and the significance we assign to everything from casual conversations to catastrophic events, determines whether we live a life smothered in illusions or a life of expanding freedom.

In this book Dr. Parrish takes us on a walk. The flagstones we tread are concepts, each seemingly unique and separate. Yet because we are walking with a teacher who sees not only the path but its destination, we begin to notice that each concept leads gently and logically to the next. Subjects such as quantum physics, mysticism, cognitive psychology, and existential philosophy are interspersed with fascinating discussions of the effects of humor and the dynamics of loving relationships. The result is that we discover that we already know more than we believed we did. We simply had not noticed the connections between the ideas already within our mind.

While introducing us to what some may think of as exotic and daunting thought systems, Dr. Parrish opens our eyes to the truth that all knowledge is already ours, although still locked, perhaps, beneath layers of unconsciousness. We further discover that there are multiple ways this knowledge can be drawn on and utilized now, today, in highly transformative ways. The chaotic landscape of our lives, this perplexing and often cruel world we seem hopelessly trapped in, can be transmuted into a comfortable and softly embracing Home. In short, projection can be put to Divine use.

By way of example of how Dr. Parrish takes seemingly separate concepts and weaves them into a usable whole, consider the following six:

1. Nothing I see means anything.
2. Everything I see is a projection.
3. My body's eyes are used to see what I outwardly project as separate from me.
4. Within the Divine nothing is seen because everything is experienced as One in the eternal present. Since nothing is separate, eyes are not needed.
5. It is therefore what is *not* seen that embraces us all in love, in happiness, and in an unalterable Truth.
6. Once acknowledged, this Truth is reflected in our projections.

Connected in this way, these concepts lead us to an ancient and familiar conclusion: *I do not have to figure anything out by myself. I can safely resign as my own teacher, lean instead on the sustaining Infinite, and put all things in God's hands*. This is a teaching contained in all the great scriptures and mystical writings as well as in problem-specific approaches such as the twelve steps of Alcoholic Anonymous, and it has unlimited application.

David Parrish's *Nothing I See Means Anything,* which may at first appear to be a fear-provoking inquiry, turns out to be a book that releases us from fear, frees us to trust our destiny, and makes accessible a familiar Voice within, a Presence that seeks only to bless us.

—Hugh Prather

Preface

This is a book I never intended to write: a book about expanded consciousness. Why? Because the content of this book is completely the opposite of what I had believed in and trusted. I had pursued a career in medicine where successful concepts were based on observation, identification, and measurement. This "scientific method" put into the language of everyday life is, "What you get is what you see." If you *don't* see it, you either are not looking in the right places or it just *doesn't* exist. As I looked around and collected experience, these concepts often were very applicable and effective, and they possessed appealing logic. These concepts of observation and measurement are traditional components of training and practice in Western medicine.

After about fifteen years of using the conventional methods of scientific thinking and research, I noticed that I would periodically feel there was more that was just outside the boundaries of my awareness. For me, there was the observed world that I could verify by measurement and replication, and then there was a metaphysical realm that defied direct observation and measurement. This unseen realm appeared to be based primarily upon presumption and faith. This unobserved area is generally regarded by Western medicine as unscientific, if not intellectually illegitimate.

The two areas felt unconnected, isolated, and persistently separate.

Then, over the ten-year period 1986 to 1996, several surprising events took place. I volunteered with a hospice during that decade and became acquainted with a patient whom I will call Harry Green. Harry had a terminal case of prostate cancer that had extended into his bones, lungs, and brain, causing much pain and disability. I had not seen him for many months when the following occurred. I was walking down the street with a friend one afternoon when I looked across to the other side and saw a person who vaguely resembled Harry Green walking toward me. I remember thinking, "Bob, what a close resemblance this man has to Harry Green." But the closer he came, the more he looked like Harry. Then it became apparent that this man *was* Harry Green. I crossed over to the other side of the street, still in a state of disbelief, and mumbled something like, "Harry, is that you?" He replied, "It sure is." I countered, "The last time I saw you, you and your family were picking out a cemetery plot." He replied, "I just decided that I am not going to have cancer any longer." I hesitated a few moments and then I said, "Harry, you just can't say you're not going to have cancer any longer." With unusual conviction and composure, he replied, "Well, I just did." And it was true that Harry's cancer had come about with shocking swiftness. I'd always been certain that malignant tumors are governed by fixed laws of biology, chemistry, and physics, and do not change with a whim of the patient's mind. Yet I had to wonder, for Harry's consciousness had undergone a massive change.

The second event that propelled me toward writing this book came about by way of a request from the Veteran's Administration. In 1995, the Veteran's Administration requested that I evaluate some 120 POWs to determine if they possessed service-connected posttraumatic stress disorders that would merit compensation. At this point, it is of some interest to note that in spite of the trauma of imprisonment and the related suffering, not one of the POWs had ever independently requested a medical exam for service-connected posttraumatic stress disorders to determine disability. I would regard this reticence as an aspect of their PTSD, and it is all the more remarkable when you consider that, in the United States, we live in a culture that encourages and rewards attempts to gain compensation awards (which not infrequently involve extensive punitive damages). My evaluations revealed that approximately 75 percent of these POWs had some degree of posttraumatic stress disorder. About 24 percent showed negligible signs of posttraumatic stress disorder, and the remaining 1 percent (two POWs) showed what might be called supernormal emotional health and, not surprisingly, above average physical health.

One veteran, whom I will call Steve Sloan, was astonishingly well adjusted compared to anyone and everyone. Sloan was a B-17 captain during WW II and was the last one to leave his crippled plane. The navigator, bombardier, tail gunner, and wounded pilot had already parachuted out. He was captured in occupied France and placed in a POW camp, where he remained for seven months. Shortly before the Allied liberation of France, Sloan escaped to Allied lines, narrowly escaping being killed by friendly machine gun fire. Steve Sloan's wife stated that prior to his entry into the service he had been a very friendly, likeable, and outgoing person, but she was shocked and unbelieving when she met the person Steve Sloan had become after the war ended. She expected him, at the very least, to be resentful. He was just the opposite of what she'd feared. Somehow, a transformation had taken place. He was now extraordinarily caring, tolerant, and peaceful. Further, this change was fixed and ongoing. When I spoke to him, these characteristics were preeminent. He possessed a quiet confidence that life would proceed in a harmonious way. There was a noticeable absence of any sense of tension, scarcity, fear, or guilt. He had a quiet self-esteem that was not based on the opinions of others. He expressed the conviction that what others thought of him was none of his business. Further, he was primarily focused on the present moment, unlike most of us who have one foot firmly planted in the past and the other foot firmly planted in the future.

How was it that these people transcended their circumstances to reach much more desirable and higher levels of consciousness? They had obviously learned things that most of us had not.

Although every physician has seen people similar to the ones I've just described, great healers are just too rare to examine and study as a category. Consequently, we really do not know their secrets. I believe that what makes such people extraordinary is not something they do; rather it is something they are!

Taking the opportunity to explore the extent of consciousness is not something that most of us do. Some people take more advantage of this opportunity than others. They do not do more or feel more or think more. Somehow, like light passing through a crystal-clear lens, life and meaning comes through them clearer, brighter, and sharper than it does through those who choose not to explore their consciousness. Because of these enlightening experiences that were so surprising, appealing, and magnetic, I felt I had to study expanding consciousness. I did so by tracing expanding awareness through the conventional sciences and to a developed and defined metaphysical reality and a metaphysical ground.

Introduction

The brain is wider than the sky,
for—put them Side by Side
the one the other will contain
with ease—you beside.
The brain is deeper than the sea
for hold them Blue to Blue
the one the other will absorb
as sponges—buckets do.
The brain is just the weight of God
for heft them Pound for Pound
and they will differ if they do—
as Syllable for Sound.

—Emily Dickinson, No. 632, 1896

Who are we? Where are we going? What does all this mean? Can this universe be a place where we can live in harmony and peace?

We seekers of truth have examined the world from three radically different frames of reference—psychology, quantum physics, and mysticism—for recurring and shared ideas. We have sought balance and perspective in this exploration, keeping in mind a comment by P.D. Ouspensky (1997), "A religion (metaphysics) contradicting science and a science contradicting religion are equally false." We have found sufficient similarity in psychology, quantum physics, and mysticism to construct what we believe is a coherent image of reality—an image that can open the door to a new and profound examination and understanding of our universe.

In the last half of this century, our three sources have spawned a veritable flood of new ideas that contend that humans are not as they think they are, and that their world and universe are not as they see them. This information has been met with universal shock, resistance, and disbelief. Particularly shocking is the observation that true reality is exactly the *opposite* of what we believed it to be. Because the new concepts appear so radical, they connect with our present knowledge only with the greatest difficulty.

I intent to provide a journey that will explain how certain information revealed by psychology, quantum physics, and mysticism was discovered, how that information interconnects, and where it will lead.

Psychology had initially indicated that much of human behavior was directed by drives and motivations totally outside of awareness. With few exceptions, humans are destined to remain unaware of these hidden forces throughout life. Later developments in psychology found that this hidden information was not completely locked in the unconscious, but was strongly denied by *choice*—a choice often forgotten. The discovery that the mind has a choice is an optimistic development, discovered by using scientific methods to gather data that supports the surprising conclusion that much of man's knowledge about himself and his world has been totally mistaken.

I believe man has long held the erroneous view that his reality is quite separate from that of his fellow man, for he has lived his life, with or without intent, dedicated to separateness and fragmentation. The denial of man's interconnectedness has left most people with lives of narcissism, conflict, isolation, guilt, pain, and depression. In other words, what we hold to be true is projected or extended beyond ourselves, and then viewed as separate from the self.

Recent discoveries in psychology have reinforced the idea that man can become capable of identifying his beliefs, and then eliminating the false beliefs of scarcity, guilt, and fear. This process can bring about a happier, kinder, more satisfying consciousness. What we think and feel determines what we experience and perceive.

With the growth of our understanding of quantum reality, all the rules for consciousness were suddenly changed. There were no stop signs. Red lights were green, dead ends were now through streets, and solid walls had open doors. Quantum physics undeniably demonstrated that we can truly choose our level of consciousness, and that we can choose either a positive or a negative collection of beliefs. It is a sad observation to see that most of us have chosen negative beliefs, and continue to believe that we are victims of circumstance. The good news is that these circumstances are situations we have constructed, and are therefore ones we can change.

Like modern psychology, quantum physics not only states that there is another order of reality that lies beneath the reality that we ordinarily perceive, but offers proof that can be confirmed by scientific experiments. The perceived reality is the one most of us believe to be real, because it seemingly can be directly measured and experienced. It is a reality characterized by apparent chance, separateness, scarcity, unending conflict, and chaos. In fact, this old reality is inside out, upside down, and backwards. The newly discovered order of reality underlying the universe speaks of wholeness, harmony, timelessness, nonmaterial intelligence, beauty, truth, and good.

What we observe in the world is the projection of our own errors of thought. Therefore, we can fail to understand only what is not understandable. By choosing to let go of these illusions, we allow the true and understandable to be known.

The prominent quantum physicist and Nobel Prize winner Paul Davies said in 1980, "These statements [about quantum reality] are so stunning that most scientists lead a sort of dual life, accepting them in the laboratory, but rejecting them without thought in daily life." Valuable concepts that are clearly optimistic and peaceful continue to be rejected and isolated from mainstream thinking. Quantum physics has made the astonishing discovery that in certain experiments involving protons, the consciousness or mind-set of the observer will determine outcome and reality. Can it be that we all choose not only our awareness, but also our reality? For most of us, this idea sounds alien and strange indeed. If we seriously assess the degree of disharmony that we see in ourselves and our world, we are forced to conclude that if we have the choice of acquiring a more harmonious view, a quantum view, this choice should be sought.

Physicist and Nobel Prize winner in physics David Bohm, in his 1991 book *Changing Consciousness*, made this illuminating remark on our need to move toward a more enlightened consciousness:

> Conflict has come from our thought, and it has evolved over the whole period of civilization. Thought has developed in

> such a way that it has an intrinsic disposition to divide things up...We could sum up by saying that we've got to look at thought. In this process, we will come upon some more subtle quality of the mind that will begin to awaken and that can spread. This subtle quality will show up collectively in a sense of impersonal fellowship that generates trust, and in the intellect as a thinking together that is free of the general pressure toward self-deception that we now feel. This can be the germ of a radically new kind of culture.

A surprising number of modern sages and mystics have supported science and in particular, physics. They have provided spiritual and metaphysical views that support the information coming from quantum physics. Finally, it has become apparent to many that metaphysics and mysticism are extending consciousness far beyond the scope of psychology and quantum physics.

In the following pages, I intend to identify bridges of information that can lead to an unfolding of this unbelievably exciting, optimistic, and enriching new consciousness. Of particular importance is the identification of projection as an early psychomarker of expanding consciousness. I invite you to learn valuable insights from physics, psychology, and mysticism that may slowly but surely alter your self and worldview. The messages from these three systems are interconnected and powerful, yet harmonious. They assert that man has the ability not only to discern the truth of his existence, but also to recognize that this truth represents his own underlying reality—a reality of unity, love, and timelessness.

I

Freudian Psychoanalytic Psychology

Hidden Desires

In the beginning, there was desire, which was the first seed of the mind; sages, having meditated in their hearts, have discovered by their wisdom the connection of the existent with the non-existent.

—"The Hymn of Creation" from the *Rig Veda*

It is essential to your understanding of evolving consciousness that we consider in some depth the three fundamental disciplines that are the heart of this book. We will begin with an examination of Freud's psychology, and in particular, two of his most important discoveries: the function of projection, and the realm of man's hidden desires known as the unconscious.

Freud discovered that projection was a mental mechanism that maintains a crucial mental equilibrium. He defined projection as an unrecognized unconscious entity involving both repression and displacement. Repression takes place when feelings and ideas are expelled and withheld from conscious awareness. Displacement may be understood as the transfer of energy from the original object to a substitute or symbolic object, chosen because it possesses a less fearful, more neutral emotional status. Less invested with affective (emotional) energy, the replaced object is more acceptable to our internal censoring and passes more easily beyond the boundaries of repression to consciousness.

The mechanism of projection has an all-encompassing application throughout mental life. Projection operates in psychosis, neurosis, and dream work, as well as in normal behavior. If one doubts the prevalence of projection, he needs only to consider the prevalence of that very common and universal pastime, creating scapegoats. In this process, guilt is relieved by blaming other people or external circumstances for events that are primarily of one's own doing. This is known as the *not me syndrome.*

One of the things that become apparent when the mechanism of projection is examined is its pervasive existence in mental life. Historically, projection has been viewed as a function of the unconscious, where it serves to neutralize and express unacceptable strivings. Psychology, for the most part, has limited the concept of projection to psychopathology, the only exception being in childhood, when projection facilitates normal development. For example, this can be seen when creative children hold conversations with an imaginary twin, with a resulting increase in their innovative problem-solving and communication skills.

Projection can also occur when wishes and drives are non-conflictual. Here, wishes and drives are available for recall; however, recognition depends on the degree of awareness. For example, if there is focused concentration on an idea, the idea will frequently be reflected to the outside world as a physical counterpart. It has long been stated by psychologists, philosophers, and savants that a person must be careful for what he wishes because it is likely to come about. In this book, we will use the term *projection* as a process that will externalize both repressed as well as non-repressed mental content. Depending on the distortion or nondistortion of mental content, projection may result in either an accurate or inaccurate perception and interpretation of the outside world. When mental content is distorted, consciousness recedes; when mental content is undistorted, consciousness advances. It is crucial to remember that all that is projected outward is reflected back to the mind *unchanged.* This latter process is called introjection.

7

In 1898 Sigmund Freud, then an obscure neurologist, startled the world with his publication *The Interpretation of Dreams.* If there were a Richter scale for psychic tremors, Freud's discovery of the unconscious would have registered a ten. There had been two previous psychic tremors of similar proportions: in 1543 Copernicus discovered that the earth was not the center of the universe, and in 1871 Charles Darwin discovered that humans are descendants of the animals. The universe was no longer man-centered. With Freud's discovery of the unconscious, man could no longer maintain that his behavior and feelings were directly under his control.

Freud, as well as many of his contemporaries, had long suspected that behind Victorian propriety and decorum laid a vast, smoldering pool of forbidden desires and impulses pushing for release. Freudian psychology was the first science to identify the extensiveness of man's denied hidden thoughts and feelings, and then to demonstrate their massive impact upon behavior.

Science had made immense progress prior to Freud's discovery of the unconscious. This progress had been brought about by the process of breaking down each entity of study into smaller and smaller fragments. Freud vigorously adhered to this scientific model and attempted to break down, or analyze, each piece of mental life. By repeated analysis of these fragments, he hoped to find their true meaning and origin. He compared psychoanalysis to working on a huge jigsaw puzzle with many colorful and detailed pieces; these pieces remain incomprehensible until they are fitted into a complete picture. Freud felt that all of human behavior was directly or indirectly influenced by repressed and often repudiated drives and wishes. He thought that these mental contents and processes were prevented from becoming conscious because of the strong defending counterforce of censorship and repression. This censorship and repression was defined as that mental mechanism whereby an idea or feeling is either withheld or expelled from consciousness, because it is considered abhorrent, threatening, or unacceptable.

Freud contended that the content of the unconscious was limited to wishes seeking fulfillment and consisted of unacceptable, primitive, sexual, aggressive, or self-sustaining drives, including murder and incest. These drives he regarded as continually striving for release in thought and behavior. This constant striving created internal conflict in the mind, between the need to contain the drives and the urge to express them. Their discharge functioned to reduce psychic tension.

The goal of this remarkable arrangement is a mental equilibrium that can be compared to that attained by an acrobat on a tightrope. The acrobat's forward progress is accompanied by unbalancing shifts of body weight that are continually countered by shifts in the balance pole. Mental equilibrium also prevents the mind and body from being overrun by unbalanced inner tensions seeking release. An American physiologist, Walter Bradford Cannon, was the first to conceive the immensely helpful concept known as homeostasis (Cannon's law) in 1932. He stated: "Organisms composed of material which is characterized by the utmost inconsistency and unsteadiness have somehow learned the methods of maintaining constancy and keeping steady in the presence of circumstances, which might reasonably be expected to prove profoundly disturbing."

In formulating his concept of equilibrium, Cannon did not mean something set and unchanging. On the contrary, he saw mind and body as extremely flexible and changeable. Man's equilibrium is continuously disturbed, but is also continuously reestablished. Freud's formulated concept of the unconscious superbly fit this homoeostatic principle. This arrangement allowed for the discharge of repressed instinctual wishes in a way that preserved the intactness of the self. The maintenance of manageable tension, as well as the discharge and gratification of repressed forces, is the ultimate aim of this energy. Freud noticed that the energy behind repressed mental processes was highly displaceable. He identified a number of unconscious mechanisms that would allow for discharge and gratification, while maintaining this psychic equilibrium. This energy became displaced and most often projected as it moved closer to discharge. As the method of discharge becomes more effective, efficient, and closer to awareness, the constraints of time, energy, and logic become more prominent. For example, a person who has a strong religious or ethical prohibition against pornography may find themselves on a censoring panel reviewing books or films for pornographic content.

Science shows organisms composed of material are characterized by inconsistency and unsteadiness, but have learned methods of maintaining constancy and keeping steady in the presence of circumstances. The human mind functions in much the same way. The maintenance of manageable tension, as well as the discharge and gratification of repressed forces, is the ultimate aim of this energy.

The defense mechanism of projection is central to the process of maintaining mental equilibrium in both Freudian and general psychology. This defense mechanism of projection has a much wider application than was previously considered. Some of the clinical phenomena that commonly provide examples of projection are the content of dreams, the phenomena of transference and countertransference, and parapraxes. These terms will be explained in the following paragraphs.

Parapraxes, or Freudian slips of the tongue, are part of normal behavior and provide one of the most fascinating examples of projection. Again, they represent conflict between forces striving for release and those forces opposing it. Here the forbidden wish has been repressed and remains outside of awareness, but its derivative becomes a distorted expression no longer counter to the opposing will. The expression is intended as camouflage and allows one to avoid any responsibility for the content of the slip. This projects ownership for the slip to an unknown entity; e.g., "the person who said that was not me."

The Freudian concepts of *transference* and *countertransference* are further expressions of the projection process. Transference is a process in which the patient unconsciously and inappropriately projects to those around him patterns of behavior and feelings that were associated with significant figures from childhood.

The more the therapist comes to resemble identities from the patient's past, the greater the intensity of the transference process. In psychoanalytic therapy, development of adequate transference intensity and its resolution are the principal means of reducing inner mental conflicts. An example of transference would be the reaction of shame and guilt by an employee to an employer's helpful suggestion. This can represent the employee's automatic response based on childhood experiences at the hands of a highly critical parent.

Another example that vividly demonstrates projective mechanisms is the dream. Freud uncovered the fact that the dream expresses the desires of the dreamer carried out by people and situations within the dream. While one might consider these expressions of repression, censorship, and displacement, the mechanism of projection can again be seen operating throughout the dream, since all expressions in the dream represent the dreamer.

Many of the early psychoanalytic cases graphically demonstrate the extent and the strength of the projective mechanism. The earliest recorded case of psychoanalytic treatment was that of "Anna O." She was, in fact, Bertha Pappenheim (1859–1936), who later became famous as the founder of the social work movement. Anna O. was twenty-one years old when she began in the fall of 1880 to show signs of a severe psychological disturbance.

Yes. This will suit my patients very well. The truth is, I am a psychiatrist. Born in the early summer of 1884, just outside Vienna. I remember how my stepfather...

PROF. FREUD DISCOVERS FREE ASSOCIATION WHILE VISITING A VIENNA FURNITURE MART

The immediate occasion for Anna O.'s illness was her severe mental and physical exhaustion after nursing her ailing father for several months. She was devoted to her father, who succumbed to a terminal illness the following spring. Joseph Breuer, a close associate of Freud, treated Anna O. from December 1880 until June 1882. During this time, she exhibited the most remarkable series of symptoms, which included rigid paralysis, loss of sensation in the extremities on the right side of her body, and the loss of her ability to speak German while retaining a perfect command of English, the language in which Breuer had to conduct most of her treatment.

Breuer recognized the case as one of hysterical personality. Breuer used a form of hypnosis that allowed Anna O. to uncover the hidden and terrifying thoughts that had led to her hysterical symptoms. The healing process released pent-up emotional energy and, to Breuer's amazement, relieved the patient of her symptoms, including her often agitated state of mind. By tracing back each symptom to its precipitating thought in reverse chronological order, Breuer was able to restore the patient to a fairly normal life.

Each of Anna O.'s symptoms proved to be rooted in specific psychic conflicts and moments of fright, first experienced during the course of nursing her father. The very last of Anna O.'s symptoms, the paralysis of her right arm, was finally relieved after she reported a previous hallucination where she saw a large black snake while sitting by her father's bedside. While under hypnosis, she recalled herself attempting to ward off the snake using her right arm. However, it was paralyzed. To allay her fright, she tried to say a prayer in German but found her German language blocked—she was able to pray only in English. The recovery of this memory released pent-up guilt and anxiety, resulting in the disappearance of her paralysis. Once again, she was able to speak her native language, German.

The medical cure at this time was nothing short of breathtaking. Breuer later published an account of this case in a book coauthored with Freud titled, *Studies in Hysteria* (1895). An analysis of this case led to the discovery that while Anna O. was nursing her father, she had intense and vivid fantasies of an erotic quality that were subsequently repressed. The fantasy of the black snake represented the father's phallus, which was then psychically hidden, repressed, and displaced. Mental mechanisms prevented her from acting on these hidden forbidden impulses, resulting in the paralysis of the right arm, which had been pressed against the side of her chair most adjacent to her father's sickbed. Put another way, hidden and forbidden impulses were projected onto the body, altering its function in order to prevent discharge, thereby disguising their true origin. After the treatment removed a sufficient amount of distorted mental content, she recovered, and in later life Anna O., as Bertha Pappenheim, was then able to pursue her pioneer work in the women's movement. She wrote on a wide variety of subjects and once humorously quipped, "If there is any justice in the next life, women will make the laws and men will bear the children."

Another of Freud's cases was that of Elizabeth Von R., who had been sexually abused as a child. Elizabeth was a woman of twenty-four, who as an adult had developed an impairment of walking and posture. The upper part of her body was bent forward, and she became incapable of walking without support. In addition, she had developed areas that were inordinately sensitive to pain, a symptom known as hyperalgesia. If the hyperalgesic skin or muscles of her legs were touched, she assumed a strange expression that seemed more like pleasure than pain. Up to this point in his treatment, Freud had used a process of suggestion and hypnosis to gain access to the patient's hidden experiences and impulses. This latter process tended to contaminate the treatment by stimulating the development of an excessive passive-dependent and sexualized or erotic transference toward the therapist. Astonishingly, it was Elizabeth Von R. herself who suggested to Freud that he use her free associations to better access her repressed sexual abuse and associated emotional feelings. This process led to the relief of her disabling symptoms.

An additional, earlier case of Freud's was that of Rosalie H., a woman of twenty-three, and an accomplished singer. Part of her vocal range could not be controlled, and she experienced constriction and choking. These hysterical symptoms were related to her uncle's sexual approaches, which had badly frightened her. Again, Freud was able to rid her of these symptoms by helping her recall memories of these traumatic experiences and then instructing her to express the hidden feelings, thereby removing her distorted projections.

In these last two cases, traumatic events of childhood produced little or no effect on the child at the time, but with emerging sexual exposure and understanding, they gathered traumatic power that led to both protective and expressive symptoms. In summary, these patients had undergone traumatic experiences causing intense excitation, both stimulating and painfully revolting. The simultaneous expression of desire and censorship was the result.

The experience represented an idea or ideas that were not compatible with aware consciousness. Subsequently, the idea was unknowingly removed from consciousness. The excitation associated with this repudiated idea was displaced, or projected, onto biological pathways, resulting in hysterical manifestations or symptoms. What remains in consciousness is simply a coded symbol that is connected with the traumatic event by disguised links. When the memory of the traumatic event is brought into consciousness, a discharge of energy occurs and the choked-off feeling from the symptom complex disappears.

Carl Jung—Swiss neurologist, psychoanalytic pioneer, and also an associate of Sigmund Freud—made the following prophetic remark regarding the expansion of consciousness at the end of his life: "All the greatest and most important problems of life are fundamentally unsolvable. They can never be solved but only outgrown. This outgrowth requires a new level of consciousness to some higher order or wider interest. Through this broadening the insolvable problem its urgency and the question itself disappear."

I would like to pay my respects to psychoanalysis for its vast contributions to the science of psychology and the art of healing. While Freudian psychoanalysis as a science and treatment has had numerous detractors, it has unquestionably supplied us with many useful discoveries and concepts. Strupp et al. (1977), in discussing psychoanalytic therapy, stated, "The golden age of psychotherapy has passed. The insights gained by the early pioneers will be woven into new approaches, some of which will bear only a fleeting resemblance to the traditional model...Chief among the retained insights is the realization that lasting characterological change is rarely or never an instantaneous and delightful experience and that education, the method of all psychotherapies is a gradual, sometimes painful process that removes the very accretions of change themselves."

Freud believed that psychoanalysis would not be effective if applied to serious, chronic psychological disturbances. History has proved Freud correct in this prediction, as these disorders also are related to persistent neurochemical imbalances. Psychoanalysis still remains a valuable option if the diagnosis is psychoneurosis and the goal is reconstructive change—time and funds permitting. Unfortunately, in America, many of the characteristics that classical psychoanalysis has considered to be pathological are currently considered virtues: manipulated relationships, arrogant entitlement, narcissistic self-promotion, immediate gratification, and impulsive acting out.

The repeated division of each mental component into smaller and smaller bits enabled psychoanalysis to fit previously indecipherable information into an intelligible whole. No question about it, this was a most extraordinary achievement. However, it was this very approach, with its blessings from scientifically observed and measured reality, that has slammed the door on any further attempt by science to knowingly fit man into the overall scheme of the planet and the universe. Until the mid–twentieth century, such a unitary overview was declared scientifically off limits and illegitimate, and it found enthusiasts only in certain corners of philosophy and metaphysics.

The genius of Sigmund Freud was most rare. It incorporated a unique blend of scientific rigor, ambition, empathy, colossal energy, and creativity. He had discovered the unconscious, a scientific milestone in the process of identifying the pervasiveness of projection and its relationship to perception. In other words, he had discovered certain feelings and behaviors that represented previously unidentifiable and hidden areas of mental life. Freud, however, was unable to truly perceive the astounding and unique seamlessness that exists between all inner mental life and all outer reality. This oversight occurred because he restricted the projective-perceptive process primarily to mental illness. The recognition of the vast importance of the relationships between projection, consciousness, and underlying true reality had to wait for the arrival of quantum physics and its two unexpected and unlikely handmaidens: existential-cognitive psychology and modern mysticism.

Summary

This chapter provides an explanation of the unconscious aspect of the mind, as formulated by Sigmund Freud. The unconscious, according to Freudian psychology, serves to maintain mental efficiency and equilibrium, by covering over thoughts and feelings that are conflictual or intolerable. Whether correct or faulty, these thoughts and feelings are projected outside the self, where they are perceived as unrelated to the self. This school of psychology believed that material in the unconscious was hidden and could be recalled only through the psychoanalytic exploration of dreams and free association. There are now more effective methods of identifying and correcting hidden mental content. Importantly, psychoanalysis has revealed the extent and power of our hidden mental content and how it becomes disowned, then displaced onto the outer world. The discovery of the function of the unconscious, and more particularly the centrality and purpose of projection, would change thinking forever. It would prove to provide an important platform for the further expansion of our consciousness.

Questions and Answers

Q: When you began your study of medicine, science, metaphysics, and expanding consciousness, you chose to begin with the work of Sigmund Freud. Can you explain how you came to make that choice?

A: In my research on expanding consciousness, I looked for an abrupt emergence of a critical-mass process. In this case it was the first appearance of a methodological-technological approach to consciousness, something that was not apparent prior to the work of Sigmund Freud. When you identify a critical-mass stage, as we have with Freud's concepts of consciousness, we have available to us a vital information-bearing hologram. A hologram is a multidimensional entity where even the smallest part of the entity contains, in condensed form, all of the information necessary for a detailed and complete expression. This early stage hologram gives us a preview of what will later unfold. The preview reveals a projection-introjection-extension process encrypted in a holographic matrix. This holographic preview is exciting, but most of all immensely helpful in providing a road map illuminating the process of cause (source) and effect. You must write the question, "What is the source?" the central portion of the guiding hologram invisibly at the top of every page; if not done in this kind of project, intention and direction will most certainly be lost. Freud used his medical and neurological experience to give us an enormously valuable gift. In a bold expression of genius he was able to view consciousness as an unfolding continuum: conscious, preconscious, and unconscious. He was further able to undeniably identify massive distortions and disassociations in every part of the continuum.

Seeing that these mental missteps caused vast and pervasive suffering, he began to develop ways to diminish and remove the connected illogical feelings of anxiety and fear. This process involved the removal of a large amount of recalcitrant guilt, which allowed repressed unconscious content to be examined "by the light of day"; its validity could then be challenged. His primary intent was the removal of the hidden blocks to expanding consciousness.

Q: Freudian theory and psychoanalysis applies to the treatment of emotional disorder. More specifically, how can this relate to the expansion of "normal consciousness"?

A: The intention of psychoanalysis was the correction of mental illness; however, Freud's discoveries led others to more closely examine the mind's distortions. This additional research confirmed that distortions extended to all thought. Their conclusions indicated that this "pathology" of the mind was very extensive—involving everyone—very serious, and I might say very unnecessary. The uncovered content in the unconscious became available for the first time and provided an explanation for much pathologic and much normal behavior. Secondly, extensive observations indicated the existence of extraordinarily strong forces within the mind that direct it toward higher and higher levels of integration and consciousness. Last of all, it provided evidence that with intention and method, content could be moved from the unconscious to consciousness. In a more enlightened era, this would be material for Nobel Prize recognition!

Q: You've concluded that projection is the most central mental mechanism. Could you further explain this?

A: Projection, which arises in the unconscious, represents a composite of stimulus, desire, content, aim, and discharge. A close examination indicates that all other mental mechanisms—without exception—are variations based on the projection model. The following idea might clarify things a bit. Projection occupies inner space and then is reflected out to exterior space, where it appears on your visual screen. It has been pointed out that judgment leads to projection, which in turn leads to perception.

Q: How could the correction of projection eventually lead to higher levels of consciousness?

A: In order to understand this, one must recognize that projection has an attached mechanism called introjection. The companion mechanism is also located in the unconscious and operates to return to the mind exactly what has been placed out on the perceived screen. Most of us misperceive and believe that what we see out on the screen comes not from the inner self, but from a disconnected external cause. The returning introjective part of the process then reinforces this belief. When this occurs repeatedly, the self maintains its current level of consciousness and does not move up the consciousness ladder. The whole process is rather analogous to the action of a boomerang, where the projected material is distorted (most is) and the boomerang hits us in the posterior at the end of its return journey. So first of all, it's helpful to develop a certain intolerance for going over the same stretch of rough road again and again. Technically, you have to become more aware of what is in your unconscious. The process of greater awareness then gradually releases you from constraints of anxiety, fear, and guilt, the primary factors that keep distorted content hidden. Upon removing them, you are now in a position to ask yourself, "Do I really deserve these distortions?" This process done repeatedly will eventually replace the projection-introjection cycle with the mechanism of extension, which possesses no limiting features, leading to higher and higher levels of truth.

2

Existential Psychology

Not So Hidden Desires

We should know what our convictions are and stand for them. Upon one's own philosophy, conscious or unconscious, depends one's ultimate interpretation of fact. Therefore, it is wise to be as clear as possible about one's subjective principles. As the man is, so will be his ultimate truth.

—Carl Jung

Amid the smoking ruins of World War II, there arose a gentle but firm spirit of philosophy and psychology called existentialism.

This new ideology presented a humanistic, holistic, and unique view of humanity. It also stressed the universal role of becoming conscious and responsible for discovering and developing the meaning of our individual existence. It advocated personal choice, personal responsibility, and personal freedom. Existentialism charges traditional psychology with overstressing internal mental conflicts as well as overemphasizing social conditions as the causes of these conflicts. It further stated that traditional psychology had ignored the absence of meaning, as had been demonstrated by humanity's lack of unity and wholeness.

While many have acknowledged the value of existentialism and its unique contribution to consciousness, it failed to take hold in either America or Europe. Existentialism pointed out that this failure was due to society's collective inclination to avoid and deny individual responsibility, finding causes and solutions for personal problems by changing group behavior rather than by changing individual behavior. Perhaps the failure of existentialism to thrive resulted from its extensive theoretical and philosophical orientation, which tended to restrict it to academic circles.

We will discuss existential psychology in some detail because it provides information that can be immeasurably helpful in understanding cognition, projection, and consciousness. Historically, existential psychology was the next psychological entity that further explained the existence of hidden thoughts, feelings, and desires that determine much of human behavior. It took the unique position that this hidden information is *not* the function of unconsciousness, but is rather *intentionally* avoided and denied. It also began to dissolve the long-established scientific boundaries between the observer and the observed. Existential psychology became prominent in the 1950s, '60s, and '70s, and it continues to influence psychological thought today. Surprisingly, concepts from both cognitive psychology and quantum physics closely interface with existential thought.

Existential thinking had its roots in a movement that revolted against a view of man that had prevailed in Western thought since the Middles Ages. This view portrayed man as materialistic and separated him from his brothers, as well as from God. This concept was based on Cartesian thought (mind is separated from material) and an associated concept, scientific materialism (the final reality is here; search no further). The existential position holds that science provides a faulty framework for the way in which observed objects and events are selected, ordered, and then studied. The existential position rejects the application of technology- and physics-based methodology as the only scientifically valid way to understand and treat human behavior.

Existentialism provides a philosophical basis for a humanistic psychology that supports an equal and symmetrical relationship between subject and subject, and between subject and object. This concept holds that a person's choices define him, but most important, they do so in a transient and flexible manner. Each person is always free to redefine himself, even radically, by choosing again. Man has no identity that is given to him by a predetermined epigenetic (gradually differentiating) process or one that unfolds when conditions are right to actualize his potential. May et al. (1958) wrote: "Existentialism in short, is the endeavor to understand man by cutting below the cleavage between subject and object, which has bedeviled Western thought since shortly after the Renaissance."

This concept of interrelationships between subject and subject, and subject and object, with their essential unity, takes on new meaning, particularly in light of the unified field phenomenon discovered by quantum physics. The physicist John Wheeler had a unitary view of the world. He made the following comment in 1975, directly relating quantum mechanics to existential thought:

> The quantum principle has demolished the view we once had that the universe sits safely "out there," that we can observe what goes on it, from behind a foot-thick slab of plate glass without ourselves being involved in what goes on…(To the extent we measure it). To the degree the future of the universe is changed, we change it. We have to cross out the old word, "observer" and replace it with the new word "participator." In some strange sense, quantum principle tells us that we are dealing with a participatory universe.

The existential psychotherapist attempts to translate the physicist's view of a participatory universe directly into the treatment of his patients. In this therapy there is no subject-object split, but a mutual process. Similarly, existential psychology demolishes the view that human nature or human essence is biologically or socially determined and "that one can devise experiments that create such a selection of objectivity" (Zimbardo 1969). In every choice a person makes, his future becomes his, by every intervention, experimental or therapeutic; the individual participates with others in the invention of a new future. The normal scientific procedure involves research and treatment limited by maintained objectivity, and planned manipulation and intervention. The existentialist states that the proper study of humans represents a participatory principle that codetermines the future. This is a liberating philosophy to say the least.

The idea that man is totally responsible for the constant process of self-definition has profound implications for the healing process and the therapeutic view of what behaviors are considered healthy or unhealthy. The existential view asserts the emptiness of past history and biological determination. These conceptions place man in a position to correct cognitive distortions by knowing that he has caused his reality and that mental content has determined what he sees in the outside world. What he sees is not an effect coming from an external cause. Unfortunately, man's anxiety and fearfulness block this awareness. Here again we see not only that cognition determines projection, but also that projection gives rise to perception. The idea that inside and outside are contiguous is a central concept in existential psychology.

How a person leads his life and how he acts depends on his willingness to see his situation clearly, which includes the manner in which he sees his relationships to the world and others. Humanistic existentialism, with its concern for human welfare, attempts to support the possibility that adequate relationships between people can exist. The adequacy of relationships hinges on the clarity with which one sees that he is both free and intentional, and that behavior can only be participatory.

The existentialist views the traditional concept of consciousness as a human process in which self-awareness imposes an *illusion* of distance between oneself and the object of one's focus, causing a sense of separation from events and objects. Existential psychology sees the gulf between the self and the object of focus as a gulf of nothingness. Here consciousness possesses the quality of free intention and engagement with the world.

Consciousness cannot exist alone, but is also free of the world of things. Consciousness is always awareness—a consciousness of something—and as such is an attitude representing a knowing or cognition. Existential psychology teaches that this attitude has an intentional way of relating to the object, according to one's deepest personal myth and worldview. These attitudes representing the self suffuse all of a person's actions and are evident in the content of his projections. In effect, a person defines himself by his freely chosen personal ideology, which becomes his commitment and the organizing principle of his life.

This self-definition can be observed rather easily in people who are dedicated to a particular ideology, such as clerics, some politicians, and educators. The goal of existential psychology is to cancel our negative choices rather than attribute them to some malfunction in the self. Withdrawal from reality to a passive, self-centered position is one of the basic modes by which people deny reality. Existential psychology provides a different paradigm, which freely links humans with the world through intentionality, and with no division between them. The position taken is one in which there is really a nothingness between consciousness and its object.

This idea is very reminiscent of Einstein's concept that the universe is a unified field of nothingness. The later contributions of quantum physics identified this nothingness as a field of nonmaterial energy, information, and intelligence. When nonmaterial information becomes associated with a marked reduction in consciousness, material appears. The conception of a nothingness, or a lack of separation, between consciousness and its object becomes the basis for the assumption that human freedom exists. It is this connectedness between subject and object (where cause and effect become one) that allows the existentialist to claim responsibility and power over whatever occurs. Existentialism has primarily represented itself as a psychology of freedom, its antithesis being a determined self-deception, where man unfortunately sees himself as un-free—a victim of his biology, instincts, and history, and of the forces of nature.

Since human consciousness is intentional, every human thought and act has both a goal and a cause. The determinist locates the cause outside himself, while the existentialist believes that each action is caused by motives that he has chosen to value: "My behavior is undoubtedly caused, but it is I who have freely determined the cause for my act."

Human freedom is not synonymous with unbridled power. Human freedom is a freedom to choose, and this freedom implies limited finite alternatives. Existentialism states that each choice limits another and that a person is limited by the factuality of his body and the things around him. Thus, the resistance of other choices and the situation of the material world is the matrix from which freedom grows.

Unavoidably, the existentialist must deal with and account for the different modes of consciousness. He does this by distinguishing between the pre-reflective mode and the reflective one. The difference between the pre-reflective (passive) and reflective (active) mode is that in the reflective mode, the person chooses or intends to pay close attention to his responses and to the situation in which he finds himself.

In the pre-reflective mode, such close awareness is not exercised, but the capacity to do so remains. In the reflective mode the person observes himself, attends to the inner process, and is aware of what is taking place. This is what the existential therapist intends to bring about while practicing psychotherapy. The therapist participates in the person's choice of looking closely at himself and focusing attention on personal and interpersonal processes. In the pre-reflective mode the person is aware of others, objects, and events, but chooses not to observe and process his awareness of them. An illustration of the passive pre-reflective mode would be that of the distracted motorist, observing the exit he wishes to take off the freeway, but driving right by it due to a distracting preoccupation.

Any response to the distorted cognition, other than clarity and correction, leads to projection of distortions. To be aware that what a person is doing is the consequence of his free commitment and choosing is the goal of treatment. This process promotes responsibility, removes the need for judgment, and places a person in a position to grant the same autonomy to others.

Such a corrective process does away with one's own illusionary system of self-serving omnipotence and over-control of self and others. Put another way, the corrective process would place one in a different, more adequate, more authentic and participatory relationship with others. When one begins to understand the unity and oneness of the universe, self-deception and coercive power become less attractive. As the distance between subject and subject, and subject and object, collapses and begins to dissolve, a person can choose to be aware of, make explicit, and articulate what he wants.

Such explicit awareness can then establish the crucial connection between desires, actions, and outcomes. This spelling-out process can provide an opportunity for an individual to see the reasons for what he does and to see how his personal myths and cognitions determine him. Being aware of these cognitions opens one to the awareness of distortions, nondistortions, and the reasons—false or true—that cognitions are maintained.

Existential psychology rejects the Freudian conscious/unconscious model and replaces it with a reflective and pre-reflective model. It is their position that the latter model possesses less separateness and greater information interchange. It is presented as more purposeful and optimistic. (Needleman, 1967, 1968) It is the author's opinion that both paradigms are very useful and can coexist peacefully. Material that resides in the unconscious, pre-reflective state is very often *resistant* to access. The division between conscious and unconscious, and reflective and pre-reflective, is like an opaque glass ceiling with small micro pores connecting both sides. One can focus on the opacity or focus on the connecting micro pores. Increasing awareness of these channels decidedly reduces the time

THE STABLE ILLUSION OF LACK

and effort it takes to correct distorted thinking. This perspective pays off in a big way, for it maintains our motivation and optimism to know that a very difficult task can and must be accomplished. We would emphasize that for this process to proceed, optimism is paramount.

Those who are existentially oriented believe that a person is precisely in the state in which he should be; if he truly believed he would be better off in some other state, he would be there. A person is where he is because he, in fact, has chosen to be there. His suffering is a result of priorities, or more specifically, the choice to ignore the negative consequences of these priorities. Psychological pain results when people choose to turn on themselves, calling themselves deficient, inadequate, depressed, and ill. At this point, they regard their life as wholly predetermined and regard themselves as victimized by the outer world. This position maintains a lonely, encapsulated, and distorted control over their logical realities.

Summary

Existential therapy purports to be a mutual search for truth and correctness rather than an identification of distortions, displacements, and projections. With corrected cognitions our distortions, projections, and displacements lose their energy and attraction. When this happens, the therapist and patient can then bond together to solve the most important problem of all: How can we be equal? In this position the humanist states, we must overcome self-deception and distorted cognitions by focusing on self-actualization, authentic emotional expression, creativity, and love, for this will permit an equal and whole connection between the self and others.

Questions and Answers

Q: Existential psychology has identified an absence of meaning as evidence of humanity's lack of unity and wholeness. How much importance would you personally place on that particular concept?

A: Well, I would certainly agree that lack of meaning is a central issue, for it maintains the low state of consciousness most people now possess. Recognizing that this is a condition that leads to stultifying depression, suffering, and anguish makes it imperative to discern whether a construct of meaningfulness exists and if it does, how one can acquire it. One of the very exciting and critical things that existential psychology provides is the identification of a definitive meaningfulness hidden in the unconscious. Existential psychology then proceeds to unpack it and to provide a dialogue-based framework that can help one acquire it.

Existential psychologists, like Freudian analysts, have indicated that there is access to the unconscious. Access to the unconscious is difficult, but with focused discipline it's within one's capability, particularly when that choice is recognized. It has been said that the choice to be happy or unhappy (to see meaning or the lack of it) is one of the secrets of the universe. I would view that comment as thoroughly accurate. Unconscious mechanisms of repression tend to hold choice outside of awareness; however, that condition is correctable.

Q: Why is there such a strong inclination to locate causes for conflict and pain outside of the self?

A: There is a quality of thinking at lower levels of consciousness that uses primitive projection and introjection as the primary pathway for reduction of guilt. That process involves trying to locate a cause that then can be identified as a separated entity or a scapegoat that then can be blamed. While this approach has some appeal because it relieves some anxiety and guilt, in the end there is only so much dirt that you can sweep under the rug. After a while that line of unwanted dirt visibly occupies your perception either internally or externally, and projection is used again—serving not only to perpetuate but also to hyper-accelerate the unchanging loop connecting outer space with inner space.

Q: Existential psychology uses the concepts of pre-reflective and reflective in regard to examining consciousness. What value would you place on these concepts in regard to expanding awareness?

A: I think that the primary contribution of existential psychology to expanding awareness is its technology-oriented intention to identify hidden destructive mental agendas and their reflection in perception. This approach resulted in the appearance in the psychological arena of aspects of nonduality and unitary field theory. Existential psychology had sensed similar information in the areas of quantum physics and metaphysics and in response formulated a pre-reflective-reflective methodology in which the observer and the observed functioned as a fused unit. This development was a notable early advance in addressing expanded consciousness.

Q: Could I then regard myself as being responsible for everything that I experience in my life, both good and bad?

A: I would respond to that question with a definitive yes. Here we have bad news and some good news. The bad news is that if I don't take responsibility for looking at the content of my unconscious and move the material into awareness, things are not going to change. The good news is this is not the only choice. A more productive choice is to try to gather one's energy for an examination of this material, because it really isn't as bad as it looks. The more you examine it, the better it looks because it is not based on reason. This material is not logical; therefore, you don't really have to keep it as part of your mental content when it's expressed as unharmonious, painful perceptions of the self and experiences with the outer world. There is a certain amount of negative collective consciousness and unconsciousness supporting what we believe. However, this negativity can be circumvented by staying in the present moment and not continuing to lie to ourselves, saying that we are unloving and unlovable, when this is not who we are. We have chosen to see ourselves as unloving and unlovable; correcting this involves a shift in perception.

Q: Since existential psychology has considerable helpful information, why has it not gained more acceptance?

A: This question is particularly relevant. Existential psychology had two conflicting stages of development, and it is the later stage that contains the most helpful information. From an evolutionary standpoint, it is almost a shipwreck. The first phase was primarily a reaction to the massive genocidal tragedies of World War II, not the least of which was the Holocaust. This initial response made its appearance in isolated European academic circles and was labeled as an *authentic response* to man's inhumanity and the absence of meaning. This response was a resigned attitude comprised of hopelessness, frozen mourning, and angst. Any counter-trends toward humor or optimism were labeled as false and inauthentic. Considering the historical events of the first half of the twentieth century, this is an understandable party-line point of view. But who would want to make such an unattractive and unappealing point of view such as this public?

The second phase was more correction oriented and was able to discern that man had a choice and if he didn't like what he had chosen, he was empowered to choose again. In exercising a more enlightened choice, he could use empathy, compassion, and a sense of global unity to choose less personal, less narcissistic, and less separating attitudes. Here, process began to uncover a resonating sense of meaningfulness and was regarded as leading to ultimate authenticity. Here, man took full responsibility for removing false content that was obscured in the unconscious. This final part of the existential evaluation had appeal and power but carried the negative baggage of the first phase and, unfortunately, just ran out of steam.

3
Cognitive Systems Psychology

Consciousness Awakening

If you do not express your own original ideas, if you do not listen to your own being, you will have betrayed yourself.

—Rollo May

The 1970s and 1980s brought the unexpected emergence of a body of psychology that provided major advances in the knowledge of projection and consciousness in the ideology of consciousness. The new entity was cognitive psychology; its forerunner was rational emotive psychology. This psychology extensively examined the mind in search of specific faulty beliefs or cognitions. It demonstrated how these beliefs directly cause maladaptive behaviors and are striking expressions of the projective process.

Two men are considered originators of cognitive psychology: Albert Ellis, a psychologist and psychoanalyst, and Aaron Beck, a physician and psychoanalyst. Ellis and Beck began this novel and uniquely successful approach to psychology after an extensive search for briefer and more effective treatments.

Freud had been apologetic that his treatment took from nine to twelve months, and had certainly not expected that it would lead to even longer treatments, sometimes almost unending ones, in the hands of disciples and followers. Freud had been cognizant of the need to shorten his therapy, but was skeptical about the success of such efforts. Two of Freud's early Viennese disciples, Otto Rank and Sàndor Ferenczi, and later Alexander French, made strong efforts to shorten the process, but all failed to find effective solutions. Over the past fifty years, many other clinicians have searched diligently but fruitlessly for more effective and efficient, and less costly, forms of psychotherapy. Most of the impetus for time-limited therapy has come from behavioral therapists, the two exceptions being Ellis and Beck.

Rational emotive therapy was developed by Albert Ellis and grew out of his practice of psychoanalysis. Ellis felt that emotional disturbances are caused by illogical and irrational thinking. The patient, without being aware of it, continually indoctrinated himself with phrases and sentences that inevitably became emotions and thoughts governing behavior. The task of the therapist was to make these sentences and phrases explicit, confront the patient with his irrationality, and eventually invite him to abandon such distortions.

A patient may have the recurring thought, *I am not a likable, lovable person,* which leads to self-defeating behaviors. In this case the therapist can confront the patient, and he can challenge the truth of such a statement and the extent it is believed.

Ellis found that the irrational concepts forming the basis for psychological disturbances are integrated into the culture and find expression in early childhood. Examples of the most common irrational beliefs are:

- It is essential that a person be loved and approved of by everyone.
- One must be totally competent and adequate at all times to consider oneself worthwhile.
- Unhappiness is due to circumstances that are external and beyond a person's control.
- It is easier to avoid difficulties in living rather than to face them.

- It is necessary to find an outside authority who is wiser, stronger, and more powerful than the self.
- There is always a correct or perfect solution to all problems, and if it is not found catastrophe will follow.

Ellis believed that if these concepts remain uncritically accepted, they lead to emotional conflict and unhappiness, a position also taken by existential psychology. Ellis' approach also stressed logic, reason, and rationality. The task of treatment leads to the isolation of irrational ideas and their replacement with ideas that are more adaptive and realistic. Ellis believed that one-time confrontations do effect a change, but permanent change requires that this process be extended over one's lifetime.

The rational emotive process regards independence, autonomy, and self-direction as guideposts to rational living and provides specific steps for achieving these objectives. He conceded that severe limits to therapeutic success may arise out of a patient's early training and biology.

Rational emotive therapy avoided the slow indirectness of traditional psychoanalytic practice and sought a straightforward approach, correcting and replacing old patterns of illogical and irrational thought. In addition to recognizing and focusing on irrational thinking, rational emotive therapy contends that people have the power to choose another way of thinking, and subsequently, another way of feeling and behaving. Ellis claimed a 90 percent improvement in the patients he treated. Rational emotive therapy represents one of the most radical and direct approaches in modern psychotherapy (Ellis 1971).

Aaron Beck was a Freudian psychoanalyst during the 1960s who had a strong interest in the treatment of depression. He had become aware that psychoanalytic treatment provided only limited benefits in neurotic depression. Since there had been a contemporary emphasis on cognitive factors in neurotic disturbances—stressing brevity, explicitness, and empirical verification—Beck proceeded to develop a cognitive approach to emotional illness.

Ellis and Beck rejected the concept that man is governed by unconscious forces over which he has little control! Beck located the principal cause of psychological disturbance in a person's misconception about himself, and his faulty assumptions and irrational beliefs about reality. Beck's emphasis rested on cognitive phenomena, particularly those cognitive phenomena that are readily available to the patient's awareness. Cognitive therapy does not restrict itself to the identification of cognitive distortions within awareness, but enables a person to also identify distortions formerly outside of awareness. Cognitive systems psychology, like existential psychology, attempts to dissolve the barrier between the conscious and the unconscious mind. Cognitive therapy purports to be an extension and refinement of commonsense methods of coping with psychological problems.

Aaron Beck's major thesis is that each person views his experience in certain ways and that the meaning one imposes or projects onto the experience determines one's emotional response. For example, depression may be conceptualized in terms of a cognitive triad: a negative conception of one's self, a negative interpretation of life experiences, and a negative view of the future. These are seen as cognitive distortions.

Conversely, Beck felt that Freud's early conceptualization of depression as inverted rage is so remote from the patient's experience that it provides little practical usefulness for the clinician. Beck stated that focusing on the patient's rage is not only unproductive, but can be detrimental by drawing attention away from distorted cognitions.

Beck's goals in cognitive modification are accomplished by specialized techniques, which focus on specific problem areas. These target areas include scheduling of activities, graded task assignments, mastery and pleasure therapy, cognitive reappraisal, rehearsal, and homework assignments, together with alternative explanations for the patient's experience.

Ellis and Beck assign the therapist an active and direct role in combating the patient's symptoms and difficulties. They believe that self-defeating behaviors and self-denigrating attitudes are typical of emotionally conflicted people and are best handled by a frontal assault. Here the patient learns to play an active part in opposing trends of passive inaction and learning to refrain from wallowing in psychic misery.

Beck noted that people who are emotionally conflicted systematically interpret situations in a negative manner, even when more plausible, optimistic explanations are available. When asked to reflect on the alternative explanations, people may realize that their initial interpretation is based on an unlikely inference. At this point, they may be able to recognize that they have distorted the facts to fit their preformed negative conclusions. He noted that these individuals typically make a number of conceptual and inferential errors.

David Burns, a psychiatrist and student of Beck's, formulated the following checklist of universal cognitive distortions. When these distortions are sufficiently extensive, they are reflected in a depressive process:

- All-or-nothing thinking: You look at things in absolute, black-and-white categories.
- Overgeneralization: You view a negative event as a never-ending pattern of defeat.
- Mental filter: You dwell on the negatives and ignore the positives.
- Discounting the positives: You insist that your accomplishments or positive qualities "don't count."
- Jumping to conclusions: (1) Mind reading—you assume that people are reacting negatively to you when there is no definite evidence for this; and (2) fortune-telling—you arbitrarily predict that things will turn out badly.
- Magnification or minimization: You blow things out of proportion or you shrink their importance inappropriately.
- Emotional reasoning: You reason from how you feel: *I feel like an idiot, so I really must be one.* Or, *I don't feel like doing this, so I'll put it off.*
- Should statements: You criticize yourself or other people with *should* or *shouldn'ts. Musts, oughts,* and *have tos* are similar offenders.
- Labeling: You identify with your shortcomings. Instead of saying, I made a mistake, you tell yourself, *I'm a jerk,* or *a fool,* or *a loser.*
- Personalization and blame: You blame yourself for something you weren't entirely responsible for, or you blame other people and overlook ways that your own attitudes and behaviors might have contributed to the problem (Burns 1980).

The efficiency of cognitive and cognitive-behavioral strategies in the treatment of depression has been supported by many controlled studies. These strategies were found to be more effective than any

"I didn't see you and you didn't see me—O.K.?"

THE UNCONSCIOUS UNDER REVISION

other form of therapeutic intervention. A surprise finding in a 1977 study demonstrated the superiority of this therapy over the then standard antidepressant drug Tofranil (imipramine). Currently, the treatment of depression is most effective using both cognitive therapy and the newer antidepressants. While Beck's initial application of his treatment method was for neurotic depression, his ideas have recently been applied to major depression, anxiety, and phobias with considerable success. One of the most interesting and unexpected results has been its moderate success with the very difficult to treat personality disorders (Beck et al. 1979).

An examination of rational emotive therapy, as well as of cognitive therapy, reveals a changing attitude toward consciousness. Again, both therapies reject the concept that humans are governed by strong unconscious forces that they can do almost nothing about. Rather, they conceptualize the motivating forces as illogical thought and attitudes that are hidden and maladaptive, but yet are available to consciousness and subject to revision. Irrational ideas and feelings that appear to be situated near the periphery of consciousness will invariably move toward a more central position in awareness when they are repeatedly identified and illustrated (Parrish 1978). While I embrace a positive approach to discerning what we all hide from ourselves, I would go on record as stating that to do away with the unconscious/conscious model can deprive us of a useful mechanism for understanding the resistance we all have to knowing what we are trying to hide from ourselves. In addition, this model provides an essential framework for understanding the origin, evolution, and function of mental defense mechanisms.

Both of these contemporary psychologies have a decidedly different view of traditional consciousness, extending it into areas formerly occupied by the concept of the unconscious. Our false beliefs not only interfere with the process of perception and the interpretation of the external environment, but these beliefs also become projected to the outside—where again they are seen as separate from their source, the thinker. When we bear this in mind, it becomes more and more clear that projection is determining our perception. If we wish to experience the self and our environment in a different, happier, and more peaceful way, we must shift our perception inward. Based on what we have discussed so far, we hope that it is becoming more and more apparent that projection is an active and very central path of consciousness.

Up to this point, man has seen himself as a victim of external circumstances. This very victimization now appears to be a projection of the content of our consciousness. One of the important aspects of this new information is the existence of choice, a choice to determine what we experience. For the most part, man has felt that this has not been available. Evidence now supports the fact that it truly is available and if our choice is not a happy one, we can choose again. It has also become apparent that this choice of happy-minded thoughts needs to be done again and again, as part of an ongoing process of correction.

In comparing the kinds of irrational mental concepts outlined by Ellis and Beck, both authors list characteristics that match almost precisely, particularly in regard to content; for example, Ellis' identification of the mental misthought, is parallel to Beck's cognitive distortion involving "personalization and blame." Ellis' listing of the irrational concept, "there is always a right or perfect solution to every problem of living" is very similar to Beck's cognitive distortion found in "all-or-nothing thinking" and "should have" statements.

Cognitive therapy is the first body of psychological treatment that has been successful across a broad and ever-expanding group of mental disorders (Beck 1990). At this point, these successes have been repeatedly validated by other independent researchers.

Upon examination of cognitive psychology, several things stand out. First, it demonstrates that what we believe, distorted or undistorted, is projected outward. We then experience it in the external world as perception. This continuous process can maintain the illusion that what we perceive originates outside our mind and is independent of it; this is completely inaccurate and inconsistent with scientific observations. Second, the Freudian concept of unconscious defense mechanisms—including denial and displacement-inversion—may not provide impenetrable barriers to threatening objectionable impulses, but more properly, provide distance between these impulses and central awareness. This concept would be more consistent with the theories of cognitive psychology and closer to clinical facts than the more static concept of defense. Finally, cognitive psychology has extended the concept of the conscious mind to include behaviors and feelings that many thought were available only through dream interpretation, examination of slips of the tongue, and that central tool of psychoanalysis—free association.

A growing number of clinicians feel that the disenfranchisement of consciousness that resulted from Freudian psychology is now undergoing revision. It now appears that a great deal of thought and feeling that earlier had been regarded as unconscious is actually consciously suppressed and held just outside of central awareness by a group of common cognitive distortions. The memories and recognitions of these suppressed distortions can be made available by using cognitive behavioral techniques. The implications for restoring man's ability and right to choose happiness and tranquility are phenomenal. When you look over the aggregate list of distortions cited by Ellis and Beck, another most interesting observation becomes apparent: The pervasiveness of these distorted mechanisms in human thinking is staggering, cutting across racial, cultural, and educational boundaries.

One certainly need not have an official psychiatric diagnosis to possess a mental life that is virtually studded with such mechanisms. Most feelings and behaviors are likely directed more by distorted cognitions than by undistorted ones. Ironically, the lack of recognition that this is so is held in place by the same cognitive distortions that we have just outlined.

Further examination of these mental distortions reveals that they have a common denominator: guilt and fear. It is further apparent that they have been around since man first inhabited the planet. Two questions come up: How did they find their way into the mind in the first place? Since they are there and are so commonplace, what possible function could they serve?

As we explore these questions, and possibly their answers, it is important to recognize that if the bad news is we have collectively chosen to acquire distorted beliefs, the good news is that we can choose to replace them with less-distorted cognitions that are reassuring, happier, and more likely to provide a peaceful outcome.

Summary

Cognitive psychology has taught us that hidden feelings and thoughts caused by distorted reasoning lead to our maladaptive behaviors. Cognitive psychology, along with psychoanalytic and existential psychology, has emphasized the extensiveness of our hidden thoughts and feelings. This system has provided us with a very specific list of mental distortions that have proved to be common and universal, serving no useful purpose. This psychology has moved a giant step forward in our quest for increased consciousness.

"Well, I'm disenchanted, too. We're all disenchanted."

A FAMILY OF COGNITIVE DISTORTIONS

Questions and Answers

Q: Research has indicated that cognitive systems psychology has wider application to psychopathology than either Freudian or existential psychology. Can you explain why this has taken place?

A: Cognitive psychology focuses more attention on content as opposed to form. Extensive focus on form leaves energy attached to the past as part of a repetitive and mindless preoccupation. Attention to content can accelerate the identification of illogical processes operating in the present. This process puts one in the position to assertively challenge the validity and appropriateness of staying connected to the content of the unconscious, with its distorted effects mirrored in the present. It is also my opinion that cognitive psychology does a much more thorough job of identifying categorical misthoughts than the other two psychologies, Freudian and existential psychology.

The dimensions of cognitive psychology can be better appreciated when you regard it as an evolutionary expression that began within Freudian psychoanalysis. Here it is worthwhile to note that both authors of cognitive psychology were originally Freudian psychoanalysts.

Q: Should I use positive thinking to reason my way to better health and happiness?

A: It is my belief that you are entitled to good self-esteem by virtue of your source. What others think of you is really none of your business. This is a more effective and enlightened response to blame and guilt. All of us in the lower stages of consciousness tend to blame ourselves and others, having decided we don't deserve better. Cognitive psychology indicates that there is no reason to feel that we don't deserve better. An important step in this adjustment in perception is to recognize that we're all doing the best we can right at this time and that we are not withholding better responses and better behavior. Perhaps tomorrow or next week you'll be able to respond more appropriately. Staying in the present and accepting what is going on right now can give you a certain peacefulness that will help you hold on to happier, less conflicted feelings for longer and longer periods. A part of the question that has to do with responsibility: I don't think that responsibility should be associated with blame and guilt. I think responsibility is best achieved by present acceptance and correction when the latter appears as a harmonious choice.

Q: Cognitive psychology takes the position that the unconscious has considerable transparency, making hidden content available with focused effort. Do you think this is an accurate appraisal of the unconscious?

A: I would divide the unconscious spectrum into two levels: an upper level, the conscious as described by Freud, and a lower level I would call *the super-unconscious.* There appears to be a much greater barrier blocking material in the super-unconscious (not a Freudian term) versus the unconscious. At the level of the super-unconscious there are deep-seated feelings: fear of loss of God's love, and fear of retaliation by God as punishment for an evanescent thought of replacing Him. The latter I would regard as both impossible and unnecessary. These feelings and thoughts just do not stand up to an examination of cause and effect. Here I believe we have effects that do not have attributes of the source or cause; therefore, they must be illusory or delusional. I don't think we created ourselves as spirit, but we certainly could have designed ourselves as material objects that have become our false selves and our false reality. These false identities can now be replaced by correction of the mental error or fantasy. If we are part of God's mind, we are logically incapable of attacking our cause or source. If we maintain the illusion that we have attacked God, we will be led to punish ourselves first, so He doesn't have to do this. Here we invoke avoidance, denial, and further psychotic thought. If you don't think this is accurate, look around the world and you will see that one-half of the globe is homicidally trying to eliminate the other half. This attack occurs on an ongoing basis. When you step back and reflect, you are compelled to see that this kind of judgment and separation is plainly insane. However, correct awareness can displace this psychotic thinking by constantly invoking the cause-and-effect paradigm.

Q: David Bohm is quoted as stating that "reality is what we take to be true." Can you explain how that comment is involved in the process of expanding awareness?

A: Bohm extended his inquiries from quantum physics into psychology, art, and metaphysics. That very pervasive inquiry gave him the ability to address these questions: Who are we? Where are we going? What does this all mean? He stated that what we perceive determines what we believe, and so therefore we need to repeatedly examine our perceptions to determine their validity. Ask yourself this question: Are these perceptions helpful or unhelpful, peaceful or nonpeaceful? Then choose useful and peaceful concepts.

I think that Bohm quite correctly laid out the path to be followed. The situation is metaphorically somewhat analogous to turning on your car's satellite navigation system. The alternative is to continue guessing and stopping at every other 7-11 asking for directions. I would regard Nobel Prize laureate David Bohm as one of the few truly lucid, transformative minds of the twentieth century in regard to conceptualizing an efficient pathway to higher consciousness.

4
The Psychologies Integrated

Portals to More Forceful Fields

Reality is what we take to be true. What we take to be true is what we believe. What we believe is based upon our perceptions. What we perceive depends on what we look for. What we look for depends on what we think. What we think depends on what we perceive. What we perceive determines what we believe. What we believe determines what we take to be true. What we take to be true is our reality.

—David Bohm

I will summarize the psychological ideas and concepts that have been presented thus far in order to carry forward our discussion in the most logical manner.

Freudian psychology took the position that much of man's behavior and feelings are due to causes held outside his sphere of awareness. Freud contended that much conflicted behavior and feelings rest on drives and wishes that are not accessible to consciousness other than through dream analysis and free association. The repressed and repudiated wishes push for discharge and are expressed in ways that are disguised and therefore unrecognizable. The mental mechanism whereby feelings and wishes are kept out of awareness was called repression. The mechanism by which repressed wishes are displaced onto the body or transferred out onto the external world was termed projection.

A careful appraisal of projective mechanisms reveals a rather remarkable thing. Not only does the human mind project hidden and conflicted wishes outward, it also projects outward those desires that are conscious, acceptable, and adaptive. In both cases it would appear to the observer that they are caused by something outside the mind, or in the outer world. It has been jokingly said that a person needs to be careful of what is desired, because it will likely appear. In fact, the only way you will encounter what you wish for is to have it first originate in the mind!

It was Freud who first discovered that there are extensive hidden and distorted cognitions that determine feeling and behavior. The Freudian concept is essentially deterministic and indirectly supports the contention that man cannot determine, nor be fully responsible for, what he feels or experiences. Expressed conceptually, man's experience is a compromise between the unknown, the unknowable unconscious impulses, and the demands of reality. This kind of paradigm locks most of us into a set of distorted cognitions with little way out.

Rather than accepting the Freudian theory that distorted cognitions are lodged in the unconscious, where they are essentially unavailable for retrieval and processing, existential psychology presents the conscious mind as *seamless* with no separate division acting as a repository or storage bin for hidden thoughts and feelings. This latter concept relates to the logic that the mind cannot hide something from itself without knowing what it needs to hide, which would include not only the content of what is hidden, but also the decision to hide it in the first place. Existential thought believes that any other way of looking at consciousness is incoherent and illogical.

If a person is capable of choosing to look inside the mind and recognize distorted cognitions, and then choose to correct them, that decision has enormous implications for a more serene reality. If one looks past anxiety, fear, and guilt, one will not only identify his cognitive distortions, but will also see how these distortions cover and camouflage primary, correct cognitions of harmony and tranquility. To discover that one has a choice, and that this choice can uncover positive helpful cognitions that are then projected out into the external world where they can be perceived and re-experienced, is immensely empowering and very good news indeed!

Existential psychology has a particular and pervasive value. It interfaces with psychoanalytic theory and practice, and it provides an embracing theory for the development in cognitive psychology of a practical hands-on technique of correcting mental distortions. We will see in coming chapters that the theory of existential psychology extends out to connect with the emerging concepts of consciousness developed by quantum physics.

Cognitive psychology has shown that the mechanism of change begins with the identification of those long-held beliefs that are self-limiting, self-defeating, and erroneous. Once these beliefs are recognized, they can be subjected to a process of examination and correction. Correction of cognitive errors results in a change of feelings and behaviors with an improved concept of the self. It needs to be pointed out that this process cannot be accomplished quickly, but is an ongoing process that needs to be attended to on a daily basis—sometimes on a minute-to-minute basis.

The cognitive psychological model is based on three main principles. First, the way in which a person structures a situation determines how he feels and behaves. The determination of the specific structure results from one's beliefs or cognitions. Second, the interpretation of a potentially stressful situation results in an active, continuing process that includes successive appraisals of the external situation, and combines coping abilities, risks, costs, and gains resulting from various strategies. Third, when people judge that their vital interests are at stake—as in situations involving change, loss, or gain—they tend to make highly selective and egocentric (self-serving) decisions. This cognitive structuring of a situation is responsible for triggering a specific affect or feeling, and for mobilizing the person to action or inaction. Depending on the content of the cognitive structuring, behavioral mobilization is directed toward flight, withdrawal, approach, or attack; the associated affects are anxiety, anger, sadness, or—conversely—a show of affection. It is the humanistic contention of cognitive psychology that man can and should view himself as capable, caring, and worthy of being cared for.

Cognitive therapy was first designed to deal with depression. Here, three principle cognitive distortions are identified:

- a negative view of the self

- a predominately negative interpretation of experience
- a negative view of the future

These three distortions lead a person to anticipate that current problems will continue indefinitely, falsely locating causes outside the mind. The cognitive model of anxiety, while differing some from the model of depression, is essentially the same.

From a scientific point of view, psychopathology is on a continuum with normal cognitive, affective, and behavioral responses to life's situations, and is essentially an exaggeration of the normal. This continuity between normal and pathological experience becomes quite apparent when our minds are lucid.

Given the fact that most people, most of the time, have behaviors and attitudes that are self-defeating, it has become more and more obvious to the perceptive observer that maladaptive thought patterns, behaviors, and feelings are very pervasive in the general population. In my view, the following observations on human behavior underscore this pervasiveness.

First is the existence of a deep conviction that leads the human mind to choose to be right rather than choose to be happy. This choice persists even in the face of consequences that either serve no one or whose benefits pale in comparison to the tranquility brought about by the choice to be happy.

The second observation that I would like to cite in support of this expanded concept of psychopathology, is that of the judgmental, global pastime known as *locating a scapegoat*. Verification of the latter observation requires no elaborate scientific methodology, but can easily be verified by a brief perusal of any newspaper. The content of this medium is such that removal of pages devoted to scapegoating, would leave little more than the obituaries, crossword puzzles, and classified ads.

Any dispassionate and focused appraisal of these two examples underscores the vast extent of the cognitive distortions that humanity believes to be true. Now, if one accepts that this latter statement is essentially correct, and also that these cognitive distortions have no doubt been around since man has populated the planet, one is led back again to the questions: How did this problem arise in the first place, and what possible purpose does holding on to distorted beliefs serve? While I would like to raise these questions again for consideration, I will address them in greater detail at a later time.

A statement that was formerly held to be true, "thinking does not make it so," is actually quite incorrect, because both existential and particularly cognitive psychologies demonstrate that thinking does make it so. The earlier statement, that projection makes perception, can now be amended in a very interesting way. We must now consider that cognition, either accurate or inaccurate, determines projection, which in turn determines perception.

Whether one looks at the mind's operation in a psychoanalytic way or in a cognitive-existential way, there is no question that the mind possesses a massive storehouse of distorted, erroneous beliefs that invariably come to determine feelings and behavior. Knowledge of the extent of these inaccurate beliefs continues to be blocked by self-deception and choosing not to look. The immense benefit of having accurate, useful, and harmonious cognitions continues to be avoided and overlooked. The identification and even the correction of cognitive mechanisms do not address the issue of how they got there in the first place. Attempts to provide a causality for this vast network of false beliefs by weaving together a composite of innate biological, developmental, and environmental factors interacting with one another, *fails abysmally* to provide any satisfactory explanation. After a thorough consideration of these causal factors, one is forced to admit that one really needs to look elsewhere.

The therapeutic success of humanistic existential and cognitive psychology rests on dissolution of any separation between subject and subject, subject and object, inside and outside, and finally any definitive separation between consciousness and unconsciousness. Put another way, we no longer have an observer observing what appears to be outside him, but rather a participator participating with *all that is*. We now find ourselves at the doorstep of quantum physics.

Summary

In this chapter, we have presented an integrated psychological basis for the development of a greater awareness of mental life. Psychoanalytic, existential, and cognitive psychology have all concluded that man has universally held concepts and beliefs that are self-limiting, self-defeating, and erroneous. These beliefs are projected outward and are out-pictured, where the causality becomes disconnected from the self. Once there is a realization that these beliefs exist, we can begin to make a choice to hold on to the concepts or choose more optimistic ones.

A distinction is made between observed reality and unobserved reality. However, discovery of the true nature of unobserved reality must be left to quantum physics and metaphysics.

Questions and Answers

Q: The concept of the unconscious seems to undergo evolutionary changes as you trace it from Freudian psychoanalysis through existential psychology and then through cognitive psychology. Can you make any clarifying comments on that process?

A: There's unquestionably a change in how the unconscious is conceptualized as you follow it through the different psychologies. In Freudian psychoanalysis, the unconscious is usually unavailable except through dreams and free association, whereas in the latter two venues there is more transparency. In all cases it takes a focused effort to bring these matters into awareness and then to gain a greater and greater recognition of how they are reflected in our thoughts and feelings.

A fascinating and important shift in consciousness is seen in the area of ethics. When you track ethical considerations from Freudian psychoanalysis thru cognitive psychology, you find an amazing transformational advance. Freudian psychoanalysis has ethical considerations that are closely interwoven within its entire process, which provides effective person-specific corrections. Significantly, the corrections are linked to unconscious dysfunctional content.

Existential psychology places ethical considerations in a spotlight of awareness that is almost confrontational. This in-your-face presentation can be a strong motivational cue to address one's lack of empathy and compassion toward others as well as toward the self. Like psychoanalysis, it reads the hidden dysfunctional content of the unconscious but probes its depth more extensively (e.g., the identification of *absence of meaning*).

Finally, in cognitive psychology the ethical considerations go beyond the intrapsychic and acquire more global qualities than you find in either Freudian or existential psychology. Its identification of misthoughts comes again from the reading of unconscious distortions. So what you have now are earlier marked advances in depth coupled with a late-breaking, almost quantum, leap in span. This is an integral advance that is quite significant with a quiet but compelling presence and visibility.

Q: If I free myself of preconceived systems of thinking, can I enter a new dimension of time and space, and experience a new universe?

A: Yes, all three systems strongly support the idea that one does have a choice. One can choose to ascend in consciousness, thereby entering a more tranquil experience of time and space, or even having an experience of no need for time and space.

Q: It would appear that diminishing anxiety, fear, and guilt are part of gaining expanded consciousness. Can you comment further on that?

A: The identification and removal of anxiety, guilt, and fear are really key components for seeing things in a different way. As long as those qualities are active—and they are very active in most of us—they get projected outward and then come back introjectively to support what we are thinking and feeling. However, there is a point in higher consciousness where we begin to see that we've really misunderstood the world because we have projected our sins on it and then saw them looking back at us. So they're really going to be viewed as quite fearful. With the removal of fear, anxiety, and guilt, you begin to see that what you feared in the world is instead in your mind, by your own choice. Since you are the author of these misthoughts you may now choose again. This revelation can lead you to see the world in gentleness, for you will begin to see your mind at one with God's, which is true nonduality.

Q: The material up to this point suggests that thinking is not necessarily a powerful contribution to the truth. Is this an accurate understanding?

A: Thinking, as we usually regard it, is probably not a great contribution to truth because of the extensive distortions of thought. I would regard awareness and knowing, which I think are quite different from thinking, as vehicles that are much more helpful in raising us to higher levels of consciousness. Replacing the absence of meaning with a meaning of far greater value—and knowing that this meaning is inside waiting to be recognized—is more consistent with a truer reality and being.

Q: Could you identify the essential messages of the first four chapters?

A: I would like to direct the reader's attention back to the emergent hologram identified at the end of the chapter on Freudian psychoanalysis. This previewing hologram emerged from the process of moving content from the unconscious to consciousness. The unfolding paradigm of cause, source, and effect was illuminated. What then became rapidly visible was the logic and centrality that thoughts cannot and will not leave their source.

The reader is encouraged to apply this question to all he observes and ask: Do the effects possess only attributes of the cause? Or, does it appear that thoughts have left their source? If they have, then illusions will be displayed. This application is particularly useful when one addresses the four lower levels of consciousness cited by Ken Wilber: the physical, the biological, the mental, and the subtle. The first four levels employ projection-introjection loops that allow no advance in consciousness. The upper two, the causal and the ultimate, are the levels where projection-introjection is replaced by unlimited extension. At these levels, thinking is replaced by knowing and seeing is replaced by vision. If thoughts cannot be separated from their source, and if observed effects have attributes that are not shared by cause, are the effects real or are they illusion? The *like from like* principle definitely applies here. Unless you look upon illusions, you cannot escape from them and look beyond to truth. If you do not look, they remain projected. Guilt and fear block the examination of illusions, but they can be removed. If guilt is not completely removed, illusions will remain and resist correction. If illusions are not corrected and removed, they will keep recycling in the mind-brain-body indefinitely. It is helpful to keep in mind that illusions cannot be dangerous.

The most important question to ask would be, "What is the identity of the cause or source?"

5
Quantum Physics

The Emerging Holoverse

I'm astonished by people who want to know the universe when it's hard enough to find your way around Chinatown.

—Woody Allen

When a man sits with a pretty girl for an hour, it seems like a minute. But let him sit on a hot stove for a minute, and it's longer than any hour. That's relativity.

—Albert Einstein

Nothing to the supramental sense is really finite; it is founded on a feeling of all in each and of each in all.

—Sri Aurobindo

In this chapter, you will find some basic aspects of quantum physics that underlie its concept of consciousness. Later, we will discuss how the consciousness of quantum physics interfaces with the consciousness of the psyche, and proceed to follow where it leads.

As if man's fragile narcissism had not been sufficiently violated by previous scientific blockbusters, we now have on the horizon a scientific discovery that is causing a psychic tremor that can easily exceed all of the previously registered "tens." To put this tremor in perspective, the difference might be compared to the difference between a popgun and a hydrogen bomb. This latest assault on man's narcissism results from the discoveries of quantum physics. Niels Bohr, a Danish Nobel prize winner in physics and one of the founding fathers of quantum physics, prophetically stated in 1933, "Anyone who is not shocked by quantum theory really doesn't understand it." Quantum physics has validated that nothing, *absolutely nothing*, exists as we see it! This includes all time, all space, and all matter. These quantities are illusions brought into perception *by man himself.* Time, space, and material are not real, but are projections extending out from the observer's consciousness. Not unexpectedly, this newest scientific discovery has caused great excitement, but also massive psychological denial.

In ancient China and during the Middle Ages, there was an inclination to view the world as composed of patterns and interconnections of separate events. Incorporated in this worldview were affinities between seemingly different things and sympathetic attractions between body, soul, and the external world. The entirety of nature was viewed as a single gigantic organism in which each individual had his or her own place. Sharing in this harmony of the universe was the pathway to right thought and action. This construct generated a configuration of knowledge that was never separate from subjective values and beliefs.

As time passed, scientific discoveries—while helpful in one aspect—created a focus of attention on separation and division that eroded the greater value of wholeness. The beginning of this fragmentation came with Sir Isaac Newton's (1685) discovery of the universal law of gravitation. The universe became describable in other ways. All the matter of this planet and the balance of the universe were no longer detached, but could now be encompassed by Newton's law of gravitation.

Mathematics joined science to describe the universe in a quantitative way that had impressive predictive force. Using the scientific method, observable facts or events could be set apart and analyzed under repeatable conditions until even the most complex processes could be reduced to a collection of basic units that act predictably as the result of the forces between them.

At the end of the nineteenth century, Newtonian mechanics had become the paradigm for all other sciences. There were those who

"... yet how is it these vital facts are virtually unknown in this country today?"

Rapidly Changing Awareness

predicted that physics would make all phenomena explainable based on a small number of physical laws. Thus, the concept of the universe was transformed from that of a living organism down to that of a machine—a machine that gave small consideration to values and meaning. It was further thought that human nature could also be reduced to the function of drives and repressions that arose from the energy generated by biochemical reactions of the nervous system.

This scientific paradigm, in which all physical phenomena can be reduced to the mechanics of their elementary units, has been very successful and has led to many important discoveries. However, there gradually developed a growing dissatisfaction with the ideas of separation. These ideas were being replaced with a feeling that the whole was much greater than the sum of its parts. While analysis and reduction are helpful, twentieth-century physics has become increasingly inclined toward an integrated and unified field approach to problem solving.

Relativity and quantum theory have had a radical effect on Newtonian physics, transforming the formalism of classical physics and altering the worldview that it long held. Niels Bohr stated in 1939 that while quantum theory had discovered the essential indivisibility of nature, Werner Heisenberg's Uncertainty Principle of 1932 revealed the degree to which an observer interacts with the system he observes. Heisenberg stated that you cannot measure mass and motion simultaneously. When you measure one, that very observation decides which value will have present reality! (*Physics and Philosophy,* 1944)

In 1975, the contemporary physicist John Wheeler graphically conceptualized this new approach: "We had this old idea, that there was a universe out there, and here is man, the observer, safely protected from the universe by a six-inch slab of plate glass. Now we learn from the quantum world that even to observe so minuscule an object as an electron, we have to shatter that plate glass, we have to reach in there. So the old world observer simply has to be crossed off the books and we must put in the new term: participator. In this way we have come to realize that the universe is a *participatory universe.*"

Despite the revolutionary ideas of quantum physics—such as the relativity of time and space, and the indivisibility of matter—the old ways of thinking have persisted in man's relation to himself and nature. Many scientists continue to search for the elementary particles out of which all material could be constructed, holding that complex biology and chemistry can be reduced to physical laws.

Consciousness continues to be regarded as only an epiphenomenon, or a derivative of the physical nervous system, when current evidence supports that it goes far beyond any physical components.

Einstein himself attempted to put the observer back into the system. He took the position that the observer can perceive different things depending upon his position, but Einstein's worldview was essentially deterministic, and came under aggressive attack by quantum adherents. Quantum physics demonstrated that universal laws were not universal at all, but applied only to local regions of reality.

These local regions happened to be the ones science had studied most habitually. As a consequence, many were slow to abandon the old view and add dimensions that were unfamiliar to them. The Newtonian paradigm remains accurate, but limited. Helpful old concepts can, without conflict, be enfolded into the much larger scientific image of reality. The machine as model and paradigm remains the starting point for physics as well as for all scientific conceptualization. The impact of the Newtonian model has been so strong that it continues to dominate the orientation of the social sciences, psychology, medicine, and economics.

Departing from Newtonian physics, quantum physics has introduced several new, and in some cases, startling concepts. Quantum physics states that the universe is made of energy, space, mass, and time, and states that *all* these entities *interpenetrate each other*. Where we observe material, there is a concentration of energy and information. This prevailing state of hyper-compaction or condensed emptiness is aptly called by quantum physics superspace. This new physics says we are the creators of what we observe; further, our consciousness is the mechanism for creating what is observed in the universe!

The influence and application of quantum physics has been ever expanding with increasing penetration into the areas of philosophy and theology. Several close observers of this revolutionary new science have even commented that quantum physics is uniquely positioned to offer a more certain pathway to God than does religion! It is of particular interest to note the unusual number of pioneering quantum physicists who have had mystical inclinations. Ken Wilber (1985), in *Quantum Questions*, was moved to call these physicists "closet mystics."

Quantum theory first emerged in 1900 with the publication of a paper by the German physicist Max Planck, on radiant heat energy. Planck had discovered that this type of energy was emitted in discrete packets he called quanta. The singular word *quantum* was subsequently used to describe modern subparticle physics.

In 1905, Einstein supported this hypothesis when he successfully explained the photoelectric effect, a phenomenon in which light energy displaces electrons from a metal surface. In order to do this, Einstein conceptualized the light beam as a shower of discrete particles, later known as photons, which in turn displace electrons from the metal surface. This description of light was in complete opposition to the classical view: Light, as well as electromagnetic radiation, consisted of continuous waves that propagate in accordance with the electromagnetic theory first presented by Maxwell in 1850.

The wave nature of light had earlier been demonstrated by the now famous two-slit apparatus of Thomas Young in 1805. The particle-wave concept was later found to have a much wider application once subparticle physics was developed. Physicists did not understand why electrons could continue to revolve around a nucleus without emitting radiation, since according to the electromagnetic theory expounded by Maxwell in 1873, the charged particles radiated electromagnetic energy while moving along a curved pathway. Based on Maxwell's theory, the orbiting atomic electrons should rapidly lose energy, colliding with their nucleus, and this they did not do.

Niels Bohr proposed in 1913 that these atomic electrons have a shared energy that permits continuous revolution without loss of energy, providing they stay at fixed energy levels. When electrons do move or jump between energy levels, their electromagnetic energy is absorbed or released in discrete packets as photons; however, this electron discontinuity was not understood until 1940. It was discovered that electrons and photons can occur as either particles or waves, depending on the particular circumstances. Bohr further proposed that the waveform corresponded to a slanting or stationary wave configuration around the nucleus. Further research indicated that all subatomic particles, not only electrons, possessed wavelike patterns.

Newtonian mechanics, as well as Maxwell's law of electromagnetism, failed to apply in the area of atoms and subatomic particles. In the mid 1920s, a new schema of mechanics—quantum mechanics—was independently developed by Werner Heisenberg and Erwin Schrödinger to explain wave-particle duality.

This new wave-particle theory was spectacularly successful, explaining atomic structure, spectra, and radioactivity, as well as effects of electric and magnetic fields. Additional extensions of quantum theory were provided by Marx Born, Paul Dirac, Enrico Fermi, and others. These extensions led to acceptable explanations for a wide variety of phenomena: from nuclear structure and reactions, electro-thermal properties of solids, creation of atoms and splitting of atoms, and the prediction of antimatter, to the stability of collapsed stars.

Quantum mechanics has led to major developments in tangible products such as computer chips, electron microscopes, and lasers. Extremely sensitive and accurate experiments have confirmed the existence and accuracy of quantum laws. There have been absolutely no experiments done in the last 50 years that have contradicted quantum mechanics. This most remarkable theory has described the universe at a level of detail and precision that is totally unprecedented.

Quantum mechanics rests on a most disturbing paradox that is related to the peculiar way subparticles can assume either wave or particle properties. Photons can be generated that will produce interference and diffraction patterns that are either energy waves or wave-particle patterns. It is not possible to anticipate whether a photon will behave as a particle or a wave until we attempt to measure it. Electrons, atoms, and other subparticles are similarly unpredictable.

If an observer attempts to measure the position and the velocity of an electron simultaneously, he finds that he can measure only one attribute at a time—either the position or the momentum. The action of measuring one quality provides undetermined and uncontrollable information, obliterating information on the other quality; this phenomenon demonstrates that an electron cannot possess both motion and position at the same time.

In the microworld of particles, there is an innate fuzziness when we attempt to measure two paired yet observable quantities. Position and momentum are paired quantum attributes, and are considered to be conjugates. Niels Bohr called this the principle of complementarity. If attributes are paired (conjugates), this relationship strangely limits how accurately you can know both. If you are precise in measuring one, you will be imprecise in measuring the other. This restriction in measurement will hold for all conjugate attributes that are dynamic (any attribute that is not constant). This paired relationship limits the precision of their natural measurement, while conforming to Heisenberg's uncertainty principle—which guarantees that all experiments will encounter a blind area just large enough to hide the solution to the wave-particle paradox. What is missing is the requirement of observer participation!

Researchers have also noticed that under certain conditions, the mind-set or conscious intent of the observer can influence the measurement outcome in subparticle physics. As a consequence of the essential indivisibility of everything, the Western concept of the relationship between the whole and its parts, between the microworld and the macroworld, and between the mind's inner world and its outer world, now had an *uninterrupted continuity.*

Many physicists thought that this quantum indeterminacy, or fuzziness, rested on a deeper substrate of deterministic principles. Einstein in particular was convinced that there were hidden laws of determination. The Einstein-Podolsky-Rosen thought experiment of 1935 represented a final attempt to dispose of Bohr's quantum principle by stating that the microscopic properties of a quantum particle must be viewed against a total macroscopic context.

What Einstein meant was that somehow, the activity of subparticles should ultimately conform to local causality, rather than nonlocal causality. In spite of the fact that Einstein had made a monumental contribution to quantum physics with his theory of relativity, he just could not accept the concept that two widely separated particles conspired to give coordinated testing results when independent measurements were made!

Bohr incisively remarked that even though no direct signal can pass between particles A and B, this does not mean they cannot cooperate in their behavior. Bohr stated that the reason they act this way is because they really are not separate, but rather possess an ongoing connectedness. In 1965, John Bell considered the problem of a two-particle quantum system and generated a mathematical theorem that evaluated particles and their forces based on logical rules of measurement.

There were two crucial assumptions in Bell's theorem. The first assumption stated there was a reality where quantum objects possess all dynamic attributes in a well-detailed sense, at all times. The second assumption is called locality, or separateness. It forbids objects from instantaneously communicating information to each other when they are not at the same location. Adhering to these two assumptions, Bell was capable of constructing a strict limit on the possible degree of correlation that would result from two particle measurements made simultaneously. This concept became known as Bell's inequality principle. Quantum mechanics predicts that the level of cooperation would exceed Bell's limit. Confirmation of quantum mechanics would require a degree of cooperation between separate objects that exceeded anything associated with any real local theory. This kind of correlation would represent signaling that was faster than the speed of light. This creative and brilliant theory provided a direct test of quantum mechanics.

Two physicists from the Lawrence Livermore Laboratory, John F. Clauser and Stuart Freedman, designed an experiment in 1972 which suggested that a kind of action at a distance was taking place between particles. The experiment received criticism by some scientists, and because of certain aspects of the methodology it never received the recognition it deserved.

In December 1982, Alain Aspect and his colleagues Dalihard and Roger completed an experiment that successfully tested Bell's theorem of inequality. Their experiment carried out polarization measurements on pairs of photons moving in opposite directions. (Calcium ions were used to generate a pulsating stream of photons.) Upon completion of their twin-state photon experiments, Aspect and his colleagues were able to show that the paired photons moving in opposite directions had a connected, correlated organization. Bell's inequality theorem was definitely *violated*. These results in turn supported Bell's quantum concept of nonlocality. The concept of nonlocality states that in the subatomic world, phenomena are connected by faster-than-light signaling. It might be more appropriate to say, as Bohr did, that phenomena are not really separated in the first place; they do not float apart in space, but rest in connecting quantum foam. These issues involved a great deal more than deciding which of the contending theories about the microworld was correct, for it would determine the very conception of the universe and the nature of reality. (Additional details of this most important experiment will be covered in a later chapter.)

Prior to the advent of quantum physics, Western scientists had assumed that the world outside of us has an *independent existence*. Objects such as stars, atoms, tables, and chairs just exist "out there," whether we observe them or not. This understanding of the world proves to be very attractive because it fits into our common sense and comprehension of nature.

Even Einstein called this objective reality because he also thought that the outside world was independent of the conscious person's observations.

With the advent of quantum physics, it was determined that all objects possessed interconnectedness, and the very conscious act of measuring a quantum object could determine its attributes. The commonsense view of reality was to be changed—changed dramatically and drastically, forever!

Prior to the advent of quantum physics, classical physics (solid-state) described the universe in a material and quantitative way. Material was separated and reduced to elementary units, and with the aid of mathematics, this science became an impressive predictive force. Quantum mechanics gradually abolished the separation and fragmentation between observer and the observed and in its place provided a participatory universe with a pervasive unity, nonmaterial intelligence, and timelessness. While there are still scientists who continue to search for elementary particles from which all material could be constructed, this approach is finally changing.

Bell's theorem may prove to be the most important theory in the history of physics. It defined the limits of local and nonlocal causality and provided an experimental method to test for nonlocal causality.

Summary

The successful experiments of Clauser, and Aspect and his colleagues, proved the existence of nonlocal causality and thereby proved that reality as we know it does not exist on a subatomic level. This research was a triumph of epochal proportions. In other words, the universal laws of particle matter are not universal at all, but apply only to local regions of perception. In discovering nonlocality, quantum physics could now factually say that the universe was far more than energy, space, matter, and time. While these new ideas have been strongly resisted by economics, thc social sciences, psychology, and medicine, these sectors are now learning that the old ideas of reality arc just not as predictive and helpful as the newer concepts of wholeness that have been described here.

Questions and Answers

Q: Why have you assigned so much importance for expanding consciousness to Heisenberg's uncertainty principle?

A: The Heisenberg uncertainty principle of 1932 unlocked for the first time the door to the immensely extended consciousness of quantum reality. The solution to the wave/particle identity paradox was found to require observer participation. This function merged the observer with the observed in an inseparable fusion, thereby indicating the vast interaction between them. These two positions had previously been regarded as strictly autonomous and separate. Stated another way, the consciousness of the observer was discovered to profoundly determine the perception and the conscious positionality of the observed. Abruptly, where there had been only measurable linearity and locality, there was now an almost unbelievable additional reality paradigm: non-measurable nonlinearity and nonlocality. Where there previously had been an *undisputed duality,* there now was the emergence of the underlying essence: nonduality, or oneness. Man's consciousness had been provided the opportunity to replace his shared "absence of meaning" with a "presence of meaning," where awareness recognizes that its content is the essence of uninterrupted oneness with all that is. Upon a very cursory examination, you can easily see that this additional view of reality is an immeasurable improvement over the prevailing view that we are trapped in the hell of an observed material world of chaos and unremitting disintegration.

Q: Bell's inequality principle of 1965 is regarded as a very significant theorem in quantum physics. Is this an example of a major expansion of consciousness?

A: Several renowned quantum physicists have called it the most important scientific theorem in the history of the world. Upon examination of all the scientific theorems that I am aware of, I would be inclined to agree. I would regard it as a landmark example of expanding consciousness. It came out of the rapidly extending scientific collective consciousness of the early twentieth century. However, it abruptly burst upon the scene of quantum physics, functioning as a radiant and organizing hologram. This theorem set definable limits for rapidity of information transfer and laid the foundation for the subsequent experiments that demonstrated that information could be transferred instantaneously.

The outcome of these experiments violated Bell's theorem, demonstrating the existence of a reality where nonlocality and nonlinearity abide. This breakthrough further collapsed any serious contention that observer/observed duality expressed any aspect of ultimate reality. At this point one might ask why it took thirty-three years after Heisenberg's uncertainty principle was presented for someone—Bell in this case—to generate an inequality principle. When you examine the inequality principle, this theorem was not generated by primary processing of information (for the pertinent information was not available), it had to wait for the development of an intuitively driven awareness that provided the knowledge that nonlinearity and nonlocality were a certainty of quantum reality. He was then able to retrospectively create the necessary theorem.

Without the awareness of this certainty, I don't believe an effective inequality theorem would have arisen. This realization very decidedly helps to identify the value of the informational perceptual and projective illusionary axis of observed reality versus the value of the axis of quantum reality characterized by vision, knowledge, truth, beauty, good, love, and oneness with infinite creation. You might note that the latter reality axis does not contain the following unappealing characteristics: anxiety, fear, loss, guilt, pain, or death.

6

Quantum Physics

Realms of Reality

I imagine that whenever a mind perceives a mathematical idea, it makes contact with Plato's world of mathematical concepts ... When mathematicians communicate, this is made possible by each one having a direct route to truth...because each is directly in contact with the same externally existing Platonic world! All the information was there all the time. It was just a matter of putting things together and "seeing" the answer...

In their greatest works, mathematicians are revealing eternal truths that have some kind of prior ethereal existence.

—Roger Penrose

In this chapter we will trace the fascinating progress of quantum mechanical theory as it zigged and zagged but steadily advanced during the last half of the twentieth century. It is one of the great ironies of that century that as reductionist biologists were slowly trying to purge all mention of consciousness from their understanding of neurophysiologic processes, physicists were at the same time uncovering compelling evidence that consciousness is not only necessary, but is integral to our understanding of the physical universe.

Albert Einstein had earlier maintained that the status of the outer world was independent of conscious observation. However, in the 1930s Niels Bohr disputed this commonsense view of reality with his Copenhagen interpretation. In this interpretation, Bohr expressed his conviction that it was meaningless to assign a complete set of attributes to a quantum object before measuring it, because the very act of measuring it changed it.

Aspect's polarization experiment demonstrated that one cannot determine what polarization (negative or positive) a photon has prior to its measurement. Further, if we measure the momentum, we will get a particle in motion, but not a position, and vice versa.

Briefly stated, a phenomenon remains a phenomenon until it is measured; this act of measurement promotes the phenomenon to reality. The philosophy that the world's reality is based in conscious observation appears alien to us, since the world behaves as if it possessed an independent existence. When we consider quantum facts, this latter impression begins to look unsustainable.

Experiments to test Bell's theorem have been conducted repeatedly; the results have been consistent and his theorem remains proved. Einstein's earlier impossible proposition has been proved to be correct.

Einstein, Rosen, and Podolsky, in 1935, constructed an argument they felt was impossible to prove. They proposed through errorless mathematical reasoning that if quantum theory were correct, then a change in the spin of one particle in a two-particle system would affect its twin simultaneously, even if the two had been widely separated in the meantime. In this case, a signal between particles would have to travel faster than the speed of light if instantaneous changes were to occur between particles.

Prior to the experiments of Clauser in 1972 and Aspect et al. in 1982, which proved that instantaneous changes in widely separated systems *did* occur, no one was ready to abandon Einstein's tenet that nothing traveled faster than the speed of light. The implication of Bell's theorem, even for physicists, was almost unthinkable. As a result of the Clauser and Aspect experiments, it must be concluded that two particles that were once in contact, separated even to the ends of the universe, can change instantaneously when a change in one of them occurs! Ultimately, the entire universe (with all its "particles," including those constituting human beings, their laboratories, observing instruments, etc.) has to be understood as a single, undivided whole, in which analysis into independently existent parts has no fundamental basis. Bohm further stated that it was his belief that each part of the universe was holographic and contained all the information presented in the entire cosmos!

The hologram is a specially constructed photo image that, when illuminated by a laser beam, appears in a three-dimensional presentation. The most unusual feature of a hologram is that exposure of only a part of it to coherent light provides an image of the entire hologram. Bohm became convinced that this principle extends to the entire universe and gave it the term *holoverse*. His theory rests on concepts that flow from modern physics that state that the world is not constructed from individual pieces, but rather is an indivisible whole. The older classical view of bits and pieces has been replaced with the concept of pattern, process, and interrelatedness. Since Bohm is regarded as one of the twentieth century's most prominent theoretical physicists, his ideas cannot be readily dismissed.

If we are part of a holoverse, then we should have holographic qualities that would enable us to comprehend such a universe. Karl Pribram, a Stanford University neurophysiologist, provided an answer to this theory in 1979 in his radical proposal that the hologram is the model for brain functioning. He has stated that the brain is the photographic plate upon which information in the universe is encoded. When we combine the propositions of Bohm and Pribram, a *new* paradigm of man is generated. We use a brain that encodes information holographically, representing a hologram that is part of the much larger hologram of the universe. Pribram's concepts are based on the research of the neurophysiologists Karl Lashley in 1950 and John Lorber in 1970, who found that information encoded in the brain is definitely nonlocal, and holographically encoded.

Roger Penrose, an Oxford mathematician and physicist, and Stuart Hammeroff, an American anesthesiologist, have discovered a structure of microtubules in the brain; the existence of this structure would support quantum mechanical functioning, thereby lending support for Pribram's concept of a holographic brain. Ordinarily we perceive the world as a collection of isolated parts that appear unrelated and disconnected. Bohm and others insist that this is an illusion and a distortion of the underlying unity that is the essential quality of the universe.

According to Bohm, the unity of the universe is enfolded as an expression of implicate order, which has already been described by physics in the motion of electron beams, electromagnetic waves, and sound waves, as well as other forms of motion. The behavior of all these forms of movement makes up the implicate order in nature.

Bohm used a simple example to illustrate hidden or enfolded order not apparent to the senses. Imagine two concentric glass cylinders, one inside the other, with a viscous fluid such as glycerin between them. The inner cylinder can be rotated slowly so that no diffusion in the glycerin occurs. One can now introduce a drop of insoluble black ink into the glycerin, and begin to rotate the inner cylinder slowly. Gradually the black drop would be drawn out into a thin thread and eventually become invisible. Rotating the apparatus in the reverse direction causes the drop of black ink to gradually reconstitute itself, becoming visible again. The drop of ink first became enfolded, invisible to the eye; however, at this point it was not a part of the unfolded reality that we could recognize, yet it was present in an implicate manner. Reversing the direction of rotation then allowed the process to be perceived by our senses, placing it in the explicate order. Bohm has further commented that the vehicle for the implicate order is the holomovement, which represents the undivided whole.

This process, first described by Bohm, explains how a quantum of energy can be displayed as either a particle or a wave. Many physicists believe that the foregoing concepts follow logically from relativity and quantum mechanics and are not daydreams or metaphors for how the universe functions. Their contention is that quantum theory embraces not only the microparticle world of electrons, protons, and quarks, but also the macroparticle world of tables and chairs, animals and humans, stars and galaxies!

Bohm's conviction is that we are in the middle of such phenomena by virtue of the time-space continuum, enfolded holographically in the universe, or holoverse. This conclusion is supported by further conclusions that follow from the arguments of Einstein, Rosen, and Podolsky, in which time and space, described as appearing to be separated, are noncausally and nonlocally related. Bohm assigned the separated moments to the explicate order and assigned the unseparated moments to the unobserved, or implicate, order.

Bohm further postulated a superimplicate order—a much more subtle and complex order—that modified the implicate order, organizing it along nonlinear and nonlocal lines. Bohm's conceptualization of this superimplicate order, as well as the associated concept of the superquantum potential, allowed him to reach his final, deepest interpretation and solution of the paradox known as Schrödinger's particle/wave equation. Erwin Schrödinger, an Austrian quantum physicist and Nobel Prize laureate, devised an equation which predicts that at any moment where a particle exists, there will be an indefinite number of possibilities extending into the future and an indefinite number of possibilities extending into the past. Schrödinger's equation has led to the speculation that there may be multiple worlds existing side by side. Because quantum particles exhibit complementarity (having the properties of both a particle and a wave), this equation is known as the wave function of the particle.

This is where the puzzle of indeterminism steps in. In certain circumstances, Schrödinger's wave function predicts the behavior of a given particle up to a point and then describes two equally probable outcomes for the same particle. On paper, as well as in observation, no reason can be found for the particle's varying behavior. The question therefore seems to have entered a kind of schizophrenic state in which it cannot decide which outcome to choose. Quantum theory does not deal with individual events. Given an individual particle, Schrödinger's wave function cannot determine where it will strike the screen but can only predict where an ensemble, or group, of particles will strike.

No example better illustrates the strangeness of this apparent connection between the observer and the observed than a now famous thought experiment known as Schrödinger's cat. In this imaginary experiment, Schrödinger pointed out that if one sealed a cat in a room with a flask of poison gas, set to break if a Geiger counter detected the radioactive decay of an atom (that has a 50 percent chance of decaying), one has no way of knowing if the cat survived the experiment unless one actually opened the sealed room to observe the results. Shroedinger regarded this paradox as representing a form of wave mechanics that was not an acceptable expression of reality. In surprisingly sharp contrast other physicists have interpreted quantum mechanics in a much different manner, where the cat inhabits two simultaneous, non-interacting but equally real worlds.

In other words, if a physicist tries to mathematically represent what is going on in the sealed room, the equations conclude only that both outcomes, both the living and the dead cat, are present in equal proportions. The fuzziness of both the quantum event and the ultimate fate of the cat are dispelled only when an observer enters the picture and perceives one of the two mutually incompatible outcomes. A little thought reveals that the dilemma posed by Schrödinger's experiment does not end with the cat, but extends fur-

Schrödinger's cat. At the end of the experiment the equation predicts that the cat will be both living *and* dead in mixed proportions.

(from B.S. Dewitt, 'Quantum Mechanics and Reality', *Physics Today*) ther. To an outsider, the observer's knowledge of what she has found is in the same schizophrenic state as the cat's fate until the observer has communicated that knowledge to the outsider. In short, the actual fate of the cat seems to exist only in the ever-expanding network of communicating observers. Put another way, the cat's fate seems to have less to do with something that exists "out there," and more to do with something that exists purely in the realm of information.

One question that follows from Schrödinger's hypothetical problem is, if the mathematics seems to imply that before an observer enters the picture, the cat should most properly be thought of as both alive and dead, what happens to one-half of the "superstate" of the cat once the other half has been perceived by an observer? The Copenhagen interpretation believes that whatever reality or potential for reality the other half possessed, exists only as a statistical reality, never to become real. However, an opposing school of thought argues that there is nothing inherent in the equations of quantum physics that allows us to logically make this assumption. According to this view, the mere fact that our act of observation has allowed only one-half of the superstate to manifest does not mean that both are not equally real at some level of reality. To say that one state is only a statistical reality is thus arbitrary, an after-the-fact assumption based on our own internal bias but not on anything inherent in the mathematics of quantum physics.

As a result, this second school of thought—known as the *many worlds hypothesis*—believes that the other half does not vanish per se, but branches off in a parallel universe that is forever unavailable to our perceptions. In other words, every reaction between a quantum event and an observer—trillions of which are occurring as the light you are reading by hits this page—splits the cosmos into a staggering number of parallel universes in which all of the ghostly states of the quantum are equally real. As might be expected, this view has some rather stupefying implications!

In an attempt to maintain the existence of an objective reality and still resolve the puzzle of the wave function, in 1961Nobel Prize-winning physicist Eugene Wigner proposed another solution. If Schrödinger's equation does represent a reality, perhaps the consciousness itself is the hidden variable that decides which outcome of an event actually occurs. Wigner points out that the paradox of Schrödinger's cat occurs only after the entry of the measurement signal into human consciousness. In other words, the paradox occurs at the point of the experiment when human observation intervenes.

According to Wigner, what quantum mechanics purports to provide is the probability of connections between subsequent apperceptions of consciousness. He asserts that it is impossible to describe quantum mechanical processes without "explicit reference to consciousness." In the dilemma of Schrödinger's cat, it is the consciousness of the observer that intervenes and triggers which of the possible outcomes is observed. Wigner also suggested that a search be made for other effects that consciousness may have upon matter.

As long as a quantum system is not observed, energy exists as both particle and wave simultaneously. However, upon observation or movement there is a collapse of the wave function into either a wave or a particle. The entry of an impression into our consciousness alters the wave function and provides a revolutionary new concept. The determining nature of consciousness has now entered the theory of physics unavoidably and unalterably. Yet there remains the question, how does this come about?

David Bohm attempted to solve the paradox of Schrödinger's equation by his formulation of the concept of the implicate order. He proposed that the particle existed more or less as an ordinary particle but that it behaves in a strange manner because it receives information through the superquantum potential, which is a wavelike information field independent of time and space. Bohm said in 1982, "The electron, in so far as it responds to a meaning in its environment, is observing the environment. It is doing exactly what human beings are doing. The measuring system, including the mind, behaves like a large integrated and closed quantum system, satisfying the Schrödinger Equation."

Recently, certain theorists have assigned consciousness to the electron and then extended it to all matter, conceptualizing that all gestalts of consciousness create their own reality (this consideration rests on the idea that consciousness structures are so integrated that they constitute a functional unit with properties that cannot be derived by adding up all the parts). The electron, in this case, exhibits a proto-intelligence where awareness is absent; this ability to create its own reality has been assigned to the superquantum potential. Returning to Schrödinger's cat and applying the concept that all gestalts of consciousness create their own reality, we would devise the following experiment: The physicist would create a three-dimensional space that includes the closed box, the laboratory, and all seen objects. Since the cat inside the box is not perceived, it is not created.

When the physicist opens the box, then the cat is created, dead or alive. The cat, too, would create a three-dimensional space, but a different one than the physicist. If the cat, with the cooperation of the physicist and all other consciousness, decides to stay alive while it is in the box, then according to its three-dimensional space (its perspective) it is alive and not in a state of suspended animation. The physicist may perceive the cat as neither dead nor alive, but that is because the physicist has not yet brought the cat into the physicist's own three-dimensional reality. Viewed in this way, the paradox disappears.

Summary

Pribram, Hammeroff, and Penrose have presented theory and information that the brain functions holographically within a holographic universe. Bohm and others have expanded the role that consciousness plays in determining reality at all levels. They further proposed that physical objects and spiritual values share a similar reality. Everybody's perceptions—the perceptions of all living creatures, of all gestalts of matter, including the electron—create an infinite dimensional space. Consciousness can construct the universe in any number of ways, with no configuration truer than another. Each has its unique focus, and each is important. All these perceptions combined build a common reality.

Questions and Answers

Q: Consciousness is an integral aspect of quantum physics. How can a further understanding of consciousness be beneficial for me?

A: Understanding the role of consciousness in quantum mechanics and the appreciation of the power of intention in this venue can provide the enlightenment to *know* that our true reality is that of unity and oneness. This realization can motivate us to go beyond our illusory ego systems, which have kept us prisoner to the Newtonian view of the world. Our fiction of a Newtonian world maintains a focus on the transient, unfulfilling investment in the measurable, unenduring aspects of external gain as a pathway to peace and happiness. This attitude locates peace and happiness outside the self, where conflicting, separate realities ensure certain failure. The most unfortunate aspect of staying associated with the ego's system of illusory reality is its inherent consciousness level, where discerning truth from falsehood is next to impossible.

Q: Perception appears to be a quality of Newtonian/particle physics; how does this interfere with acquiring the knowledge of a nondeterministic quantum reality?

A: The process of perception selects and displays the world you see. Projections of thoughts in the mind are portrayed on the screen of external reality. This process unfortunately serves to maintain the separated observer/observed paradigm, which has been proved by the experiments of quantum physics to be only an illusion of reality. Our isolated, selfish, and narcissistic egos operate on a consciousness level that involves shame, guilt, apathy, fear, desire, and anger. This process serves to block the awareness that we have a choice in what we perceive. Restated, the process of perception is a choice, not a fact. It is very difficult to become aware of how extensive separation has interfered with the function of reason. Reason remains in the self, deep outside of our awareness.

In 2001 David Hawkins, a noted research scientist in consciousness, aptly said, "Man's dilemma—now and always—has been that he misidentifies his own intellectual artifacts as reality. These artificial suppositions are merely products of an arbitrary point of perception. The inadequacy of the answers we receive is a direct consequence of the limitations implicit in the viewpoint of the questioner. If slight errors occur in the formation of questions, gross errors will occur in the answers that follow."

7
Quantum Physics

Chaos to Cosmic Order

It turns out that an eerie type of chaos can lurk just behind a façade of order and yet deep inside the chaos lurks an even eerier type of order.

—Douglas Hofstadter

As has been seen, fragmentation originates…in the fixing of insights forming our overall self-world view, which follows on our generally mechanical, routinized, and habitual modes of thought about these matters. Because primary reality goes beyond anything that can be contained in such fixed forms of measure, these insights must eventually cease to be adequate and will thus give rise to various forms of unclarity and confusion. However, when the whole field of measure is open to original and creative insight without any fixed limits or barriers, then our overall world views will cease to be rigid and the whole field of measure will come into harmony, as fragmentation within it comes to an end.

—David Bohm

Ilya Prigogine, a Belgian chemist and scientist, further confirmed the quantum mechanical concepts of interconnectedness and universal wholeness. In 1977, he was awarded the Nobel Prize for his theory of dissipative structures. Prigogine (1983) discovered and mathematically proved that while the second law of thermodynamics, entropy (the universe is gradually disintegrating), applies to the observable universe as a whole, the trend toward entropy and chaos could and would give rise to a new restructuring on a local level. In local defiance of the universal tendency toward disorganization, fluctuations can give rise to a new balance, harmony, and complexity. Prigogine pointed out that matter is not inert, but is alive and active, interacting with man. Prigogine asserts that there are parallels between the reality of the microscopic realm of the universe and the levels of ordinary human experience. He further asserts that these parallels are built into nature. The elements of a dissipative structure may act in a cooperative way to bring about a new and improved condition. This change occurs at the level of the molecule as well as at the level of human experience. Progogine (1984) has further commented that even molecules do more than continuously interact with other molecules, "but also exhibit coherent behavior suited to the needs of the parent organism."

Microscopic components reverberate and find expression at the macroscopic level. The meaning of the whole is traceable to the behavior of the part. However, the process goes beyond the behavior of the component part as new forms then emerge as the entire system evolves. The theory of dissipative structures provides for ordering to arise out of chaos, an order that Prigogine feels could not arise without chaos. Put simply, in order for things to fall together, they first must fall apart.

Small disruptions to a system tend to be dampened out, but large disruptions lead to chaos and can lead to the creation of new and greater organization. This phenomenon is sometimes referred to as "a flight to higher order." For this to occur, you have to throw a monkey wrench into the works—not a small one, but a very large monkey wrench. The HIV epidemic, for example, is a very, very large monkey wrench!

A simple but graphic analogy, representing Prigogine's theory at the level of human experience, would be that of a beautiful oak tree struck down by a lumberjack's saw. The tree would be cut into many pieces, a chaotic end to its former living state in which it provided shade and oxygen to the atmosphere. Viewed another way, its massively chaotic state can be used for a new organization: walls, flooring, and furniture.

The difficulty in accepting the comprehensiveness of Prigogine's theory lies in the traditional assumption that the world of life and no life cannot be connected except in poetry and mysticism. We are now presented with evidence to the contrary, which requires that we think in new ways about the universe. The new worldview goes far beyond the current concept of life as nothing but the blind results of molecules and their electrochemical reactions. Prigogine said that we can then see in nature what we have already seen in ourselves. This new view transcends the reduction of life to dead matter, for it sees life in matter.

Life on this planet exists because of the features of the universe. The process of evolution has been dependent on the showers of cosmic rays from outer space, which in turn have given rise to life-sustaining mutations. Not only are we materially and morphologically connected to the ecology of the universe, but the physical behavior of our bodies and everything around us is also contingent upon the large-scale features of the cosmos.

Science formerly functioned to divide and separate and was seen by many nonscientists as primarily fostering a drive for fragmentation, isolation, and destruction. Incredibly, science is now pointing in a totally opposite direction, that of cosmic unity. There are suggestions that the universe is the way it is because of how we are, and vice versa. This implication lies at the heart of the interpretation of *quantum reality* being offered by modern physicists. According to this view, a fixed, objective universe is illusory.

The physicist d'Espagnet said in 1984 that the belief in an immutable objective reality is in conflict not only with quantum theory, but also with the facts we now draw from actual reality. In 1989, Larry Dossey asked whether our shaping of the universe is directed toward the large-scale aspects of the cosmos, or toward the microscopic realms, which we cannot directly perceive. It is likely that our influence extends in both directions. Bell's theorem, which we have previously examined, suggests that conscious human activity influences the behavior of subatomic particles and ultimately shapes events in the universe. The implication that human consciousness is a factor in determining features of the real world was held in 1971 by physicist H.D. Stapp from Berkeley. A foremost proponent of Bell's theorem, Stapp believed that validation of the theorem is the most important result in the entire history of science, and that it demonstrates the effect of human consciousness at the level of the microscopic!

"Another outburst like that and I'll clear the court!"

Chaos Theory—Falling apart to fall together

One can conclude that the impact of our consciousness lies in the area of the atom and in the area of the galactic. Flowing consciousness exists on all levels of cosmic organization for consciousness affects and is affected by all events in the universe.

In view of the apparent omnidirectional flow of all information—a flow in which conscious human beings are enveloped—it would be surprising if behavior of subatomic particles were not tied to the behavior of man. If we are willing to entertain bidirectional interactions between man and the large-scale features of the cosmos, and if we are willing to examine how man's consciousness may shape events in the subatomic world, then we must conclude that day-to-day experiences are correlated with the subatomic realm. In view of the discoveries and verifications of quantum physics, this view is compelling. One should expect these associations to exist because of logical consistency, as well as the patterns of connectedness that science has already demonstrated.

It was further asserted in 1979 by Nobel physicist Eugene Wigner, that not only are life and matter connected, but physical objects and spiritual values share a similar reality. He contended that this is the only known position that can be consistent with quantum mechanics. Prigogine also has maintained that the union of physical objects and spiritual values is supported by his theory of dissipative structures.

These applications go far beyond the little effects that are being observed by a small group of physicists with expensive and complicated research apparatus. We are again reminded of Prigogine's *human physics,* so called because of the ordering principles that operate throughout the universe. Stapp's implications are similar: The oneness that is implicated in Bell's theorem envelops humans and atoms alike. To understand such a situation, we must consider that according to modern physics, we are part of the process of the universe, which by definition includes all particles—both microscopic and macroscopic. We function as a multitude of waveforms that collide and interfere, producing unending patterns of complexity. These concepts support the quantum contention that Pribram's holographic brain is enfolded within Bohm's holographic universe.

Summary

If the universe appears chaotic, then it is likely that we as constituent parts share in the chaos. However, there is a limit to this chaos, since we can make sense of things and do comprehend. We are capable of extracting harmonious processes and patterns from the universe, and we can use this information to gradually correct our distorted cognitions, thereby choosing a consciousness of unity and harmony. Corrections of cognitive distortions are entirely dependent upon the degree to which our consciousness expands to embrace human creativity on one hand, and cosmic creativity on the other.

With the recognition that the world of life and matter can be bridged, the future can only appear brighter. It can be brighter yet when we consider that Prigogine has maintained that a union of physical objects and spiritual values is supported by the extension of his theory of dissipative structures. The comments of Nobel physicist Eugene Wigner are worthy of restatement; he asserted that not only are life and matter connected, but physical objects and spiritual values share a similar reality. He contended that this is the only known position that can be consistent with quantum mechanics.

Questions and Answers

Q: Advanced theoretical physics and supporting experiments have shown that everything in the universe is dependent upon everything else; how can this idea assist us in extending our consciousness?

A: The theoretical scientists Pribram, Sheldrake, Bohm, and Prigogine vastly extended the earlier discoveries of Einstein, Bohr, Heisenberg, Schrödinger, and Bell. Their studies confirmed the quantum concepts of universal wholeness and interconnectedness. We now have a scientific foundation to inquire about the relationships between science, the universe, and consciousness as it is experienced in the mind. These insights and questions can help us detour around the artificial division between subject and object and assist us in transcending our major obstacle, the illusion of duality and its attendant limiting point of view. Once dualism is put aside, consciousness is in a position to transform its comprehension of causality. Earlier, I mentioned the crucial importance of cause and effect and the necessity of understanding whether one was observing true cause and its true effect or the illusion of false effects. When one is able to take responsibility for the consequences of his intentions and perceptions, the observer can conclude that being a victim is totally unnecessary, for nothing external can have power over you. It's no longer the events of life but our attitudes and reactions that determine whether we regard them as stressful or nonstressful.

Q: Chaos theory has been accorded the role of a central conceptualization that helps us move beyond the observed aspects of perception. Could you explain this further?

A: Prigogine conceptualized that embedded in the view of chaos, where "things were falling apart," was another observation: that things could now fall together at a higher order at *either* the explicate (observed) level or the implicate (unobserved) level.

Further investigation has led to the conclusion that these phenomena are due to organizing influences, or *attraction fields* residing in the regions of nonlinearity and nonlocality. Consequently, chaos must be regarded as a limited perception. An associated corollary of this conclusion is the fact that perception evolves with evolving consciousness, allowing us to become aware that what we thought was a world of cause is really a world of effects. This latter process provides a powerful impetus for a leap in consciousness via a marked paradigm shift.

8

Quantum Mechanics

Teleportation Realized

Vision is the art of seeing things invisible.

—Jonathan Swift

In the previous chapters, the concept of nonlocal cause and effect was seen as central to the structure of quantum physics. We are going to explore nonlocal causation in further detail because it is so crucial to the expanding concept of consciousness and our ability to determine our reality.

The photon experiments of Clauser and Aspect et al. demonstrated that nonlocal causality did indeed exist. Bell's theorem of inequality provided a mathematical limit to the degree of nonlocal correlation that can occur and still be attributed to local causality. The experiments of Clauser and Aspect exhibited this photon correlation and thus violated Bell's inequality theorem. It was not the mere fact of correlation that necessitated the concept of nonlocal connectedness, but the extent of photon correlation. Weak correlations were far more synchronized than they had any right to be. Such faster-than-light signaling (message transfers that exceed 186,000 miles per second) can only be explained by nonlocal causality. No local explanation or model of reality will suffice.

Since Bell's reasoning is accurate, invisible nonlocal connections must truly exist; their existence is clearly supported by the twin-photon research. So what justification is there for extending his conclusion beyond the validating experiments of Clauser and Aspect et al. to an everyday application, in which the reality underlying everything is nonlocal?

Quantum theory explains strong photon correlation by employing an idea called *phase entanglement*. This concept states that when quantum system A meets quantum system B, their phases become mixed up and part of A's phase (wave) goes off with B, and part of B's phase (wave) goes off with A. This phase entanglement provides for information sharing that thereafter *instantly connects* any two particles that have once interacted.

If the quantum phase connection is real—and all theorizing and testing point to the fact that it is—then it links all systems that have once interacted at some point in the past (not just twin-state photons). This connection would result in a single waveform whose most remote parts are joined in a manner that is unmediated, unmitigated, and immediate. This idea implies there is *nothing* that is not a quantum system.

The mechanism for this instant connectedness is not some invisible field that stretches from one part to the next, but a bit of each part's being lodged in the other. Each photon leaves some of its phase in the other's care, and this phase exchange connects them forever after. With the Clauser and Aspect et al. experiments, not only do we have a quantum theory that contradicts the hypothesis that the world is linked by strictly local connections, but we also now have quantum test results that support nonlocality.

There have been no small numbers of acute observers who have labeled the Aspect experiment as the "most important experiment of the century." I would have to disagree. In my view, it is *the most important experiment ever performed!* In one truly blinding flash, the curtain of time and space was ripped aside from what we had chosen to hide: the true and essential oneness of ourselves and the universe!

In 1997, at the University of Innsbruck, Austria, Anton Zeilinger and associates once again tore the curtain aside from perceived reality. These scientists destroyed bits of light in one place and made perfect replicas appear about three feet away. This feat was accomplished by transferring information about a crucial physical characteristic from the original photons. The information was received by other photons, which took on that characteristic and became replicas of the originals.

Zeilinger stated that their quantum teleportation experiments demonstrated "the transmission and reconstruction over arbitrary distances of the state of a quantum system. During teleportation, an initial photon which carries the polarization that is to be transferred and one of a pair of entangled photons are subjected to a measurement such that the second photon of the entangled pair acquired the polarization of the initial photon. This latter photon can be arbitrarily far away from the initial one. Quantum teleportation will be a critical ingredient for quantum computation networks."

This experiment was the first to demonstrate quantum teleportation—a bizarre shifting (Albert Einstein called it "spooky") of physical characteristics between nature's smallest particles, no matter how far apart they are. Another research team in Rome has reported similar success. Zeilinger predicts that we will be able to achieve teleportation between atoms and molecules within a decade. The first practical application of this Star Trekian "beam me up" technology would be quantum computers that would sling information back and forth *instantaneously*.

The experiments of Aspect et al. and Zeilinger on quantum particles, demonstrating that information is shared over vast stretches of time and space, is not simply a meaningless glitch in the fabric of the universe or an inconsequential anomaly to be glanced at briefly and then swept under the carpet. Vital questions must be asked; I believe that they can be addressed by information available *now*. Why have the Aspect-Zeilinger experiments, in particular, and quantum mechanics, in general, been met with so much relative silence, disbelief, and rejection?

One answer is that they appear to assault simple common sense. However, there is a more primary reason amply indicated by recent discoveries in psychology. It is man's persistent cognitive focus on the observed reality of measurement, separation, and chaos! Man just has not felt he is entitled to view a reality of harmony, unity, trust, and love, but has retained suppressed cognitive distortions and misthoughts of scarcity and lack that remain fixed on fear, guilt, and aggression. Mental distortions, with their projected limited perception of an observed material reality, severely restrict consciousness.

The vision of a different, much improved reality usually does not escape our ever-vigilant distorting projective process, but in this case it has. This unavoidable fact remains: The recent teleportation of light particles by Zeilinger has driven a final nail into the coffin of what is construed to be commonsense reality.

Man strongly resists seeing his observed reality as temporary or as illusionary, for doing so would seemingly deprive him of creating the permanent. Quantum reality can be seen as an attack upon observed material reality, resulting in its dissolution. If the temporary and the flawed were not real, would this be such a vast loss? The new vision for man is one of co-creation with the Divine, which is a creation of limitless and timeless harmony.

While Bell's theorem of inequality arose in quantum theory, Bell's results do not depend on the truth of quantum theory. The Clauser, Aspect, and Zeilinger experiments show that Bell's theory is violated by facts and that should the quantum theory fail, the successor theory must also violate Bell's inequality theorem in order to explain the "twin-photon state."

Our perceived local world of separateness and division in actuality rests upon a hidden and invisible reality: unrestricted, possessing faster-than-light signaling, representing a unified and unbroken wholeness! Sir James Jeans, a distinguished British astronomer, put it this way in 1983: "The stream of knowledge is heading toward a nonmechanical reality; the universe has begun to look more like a great thought than a great machine."

Since consciousness determines our experienced reality, we will examine it in greater detail. Webster's dictionary has defined the word as:

- Awareness, especially of something within the self; also the state or fact of being conscious of an external object.
- The state of being characterized by sensation, emotion, volition, and thought.
- The totality of conscious states of an individual.
- The upper level of mental life as contrasted with unconscious processes.
- Perceiving, apprehending, or noticing with a degree of controlled thought or observation.
- Capable of or marked by thought, with design or perception.
- Having one's mental faculties undulled—by sleep, faint, or stupor.

The psychiatric dictionary by Hinse and Campbell defines consciousness as synonymous with being aware. The psychoanalysts Healy, Bronner, and Bowers stated in 1930; "conscious is that part of mental life, proportionately infinitesimal, of which the individual is aware at any given time. Though consciousness is a continuum during normal waking life, its content is extremely transitory, constantly changing."

The renowned psychologist Edward Kempf stated in 1921, "The phenomena of consciousness of self or of the environment is the result of all segments reacting together, more or less vigorously as a unity, to the sensational activity of any one or several of its parts."

With all these descriptions in mind, it can be stated that consciousness is at least an awareness of ideas and feelings. More specifically, it represents the capacity to know, perceive, and arrange ideas and feelings into meaningful entities. The term is usually restricted to that part of mental life which involves the reaction and relationship between an individual and the external world of which he is aware at any point in time. Regardless of what one's specific definition of consciousness might be, two factors appear to be at the very heart of the concept: intention and attention.

Today, consciousness is a central part of modern psychology and physics and is being actively interpreted by a variety of scientific approaches. Historically, the origin of this concept has been steeped in the controversial speculations of metaphysicians, theologians, and philosophers.

During the eighteenth century, we saw an inclusion into philosophy of scientific methods and knowledge. It was at this point that the concept of consciousness began to separate from the purely speculative and became part of experimental science. This process was given momentum by James Mill, father of John Stuart Mill, with his publication in 1829 of the *Analysis of Phenomena of the Human Mind*.

Rudolf Holze furthered the process with his manuscript *Medicinische Physiologic de Scele* in 1852. This treatise attempted to explain mental life as the product of physiological processes. In 1879, Wilhelm Max Wundt authored a pioneering treatise on consciousness titled, *Outlines of Psychological Psychology*. He later founded a laboratory of experimental psychology in Leipzig and for the first time brought consciousness under the controlled investigation of science.

In this country, Sigmund Freud revolutionized the concept of consciousness with the psychoanalytic method of investigating mental processes. In Freudian psychology, consciousness is synonymous with *the conscious*. In this system, consciousness is that part of the psyche that comes in contact with the external world, reacts to it, and becomes modified by it. Consciousness is considered to be the result of the interplay of continuously striving forces within the person, as well as within the environment.

This concept was further modified by Alfred Adler, Carl Jung, and other psychoanalysts and psychologists who followed. Carl Jung was the first to suggest that beneath the familiar levels of conscious awareness and Freudian repression there lie an objective intelligence and a creative ordering that give rise to mind-consciousness. At this level, mind and matter are but two aspects of one whole and are no more separable than are form and content!

Stanislav Grof, professor of psychiatry at the Johns Hopkins University School of Medicine, in his book *Beyond the Brain* (1985) has stated, "While the traditional model of psychiatry and psychoanalysis is strictly personalistic and biographical, modern consciousness research has added new levels, realms, and dimensions and shows the human psyche as being essentially commensurate with the whole universe and all of existence." He further concluded that the existing neurophysiologic models of the brain are inadequate, and that only a holographic model can explain archetypal experiences, encounters with the collective unconscious, and other unusual phenomena experienced during altered states of consciousness.

In 1980, Dr. Kenneth Ring, a University of Connecticut psychologist, believed that these aforementioned experiences (archetypal and altered-state phenomena) are due to the shifting of a person's consciousness from one level of holographic reality to another. David Bohm has indicated that consciousness itself provides a perfect example of what he means by undivided and flowing movement.

The ebb-and-flow quality of our consciousness, while not precisely definable, can be seen as a deeper and more fundamental reality from which our thoughts and ideas unfold. Jung's concepts of the collective conscious and unconscious have received widespread acceptance, but they had no scientific explanation until the emergence of the quantum theory of interconnectedness and the holographic model of information.

In a universe where all things are intimately interconnected, all consciousness is also interconnected. There are two levels of consciousness: a personal one that is material, separate, deterministic, and related to the unfolded concept of the explicate order; and a deeper one that is nonmaterial, hidden, and not immediately accessible, which connects with all other consciousness. This enfolded, or implicate, consciousness, with its associated memories, is thought to be in resonance with upper levels of explicate consciousness.

The occurrence of synchronization also appears to support the existence of an underlying connectedness or consciousness. Events occurring in time can appear to have a synchronous appearance but cannot be accounted for on the basis of a cause-and-effect relationship. They are connected only on a deeper level where the enfolded implicated order of connectedness prevails.

In a dialogue with Renée Weber (1986), Bohm stated, "Consciousness is much more of the implicate order than it is of matter [explicate order]...Yet at a deeper level [matter and consciousness] are actually inseparable and interwoven, just as in the computer game the player and the screen are united by participation in common loops. In this view, mind and matter are two aspects of one whole and no more separable than are form and content. Deep down the consciousness of mankind is one. This is a virtual certainty because even in a vacuum, matter is one; and if we don't see this, it is because we are blinding ourselves to it."

Man has measured and formalized time and space in his perceived world in an attempt to provide order. However, in 1987 Bohm felt that the apparent separateness of consciousness and matter is an illusion that occurs only after both have unfolded into the explicate world of objects and sequential time. If there is no division between mind and matter in the implicate order, it is not surprising that explicate consciousness would show traces of the underlying implicate order in the form of observed simultaneous happenings (synchronicities). Observable synchronous phenomena are like flaws in the fabric of reality—brief fissures that allow us a glimpse of the extensive underlying unity of the universe! F. David Peat, a physicist at Queen's University in Canada, asserted in 1987 that not only do synchronicities reveal the absence of division between the physical world and our inner psychological reality, but their relative scarcity in our life also shows the extent and the degree of our separation from the deeper portion of our mind, with its infinite and astonishing potential.

According to Peat's 1987 statement, when we experience a synchronicity, what we are really experiencing is "the human mind operating for a moment, in its true order and extending throughout society and nature, moving through orders of increasing subtlety, reaching past the source of the mind and matter, into creativity itself." He further states that the apparent separateness of objects in the time-space continuum is an illusion that rests on a deeper fundamental order of unbroken wholeness.

This reasoning does not imply that individuality is lost but that it is enfolded into the underlying whole, or as Peat has stated, "the self lives on, but as one aspect of a more subtle movement that involves the order of the whole of consciousness." This whole of consciousness contains the entire biological history of not only the planet, but also the universe!

Research has found that in the world of subparticle matter, the state of consciousness of the observer determines the outcome. The observation and measurement of light collapses its duality either into a particle or into a waveform. The usual position of the observer has been replaced with one of *a participator.*

The determination of outcome that results from intention-oriented consciousness appears to be limited to the subparticle world of matter, but on closer examination it crosses over into the macroparticle world—in other words, our observed world. Information contained in a conscious thought can be conveyed to the "consciousness" of material, constituting a nonlocal resonance of meanings. Bohm felt that the electron is not only mind-like, but is a highly complex entity and not a simple structureless point. The active use of information by electrons and all subatomic particles indicates that the ability to respond to meaning is a characteristic not only of consciousness but also of all matter.

In speaking about psychokinesis, a phenomenon in which conscious thought brings about physical alteration in macroparticle matter, Bohm (1986) commented, "Meaning can serve as the link or bridge between two sides of reality." This link is indivisible in the sense that information contained in thought, which we feel on the mental side, is at the same time, on the material side, becoming neurophysiologic and biochemical with physical activity.

Bohm taught that examples of objectively active meaning can be found in physical processes, one example being the information-bearing computer chip. The measurement of this information is active since it in turn determines how electrical impulses will flow through the computer. Another example is that of subatomic particles. The standard view in physics is that quantum waves act mechanically on a particle, controlling its movement. Bohm believed that the quantum wave supplies information to the particle about its environment, enabling the particle to move on its own.

Speaking to his contention that both consciousness and matter respond to meaning, he considers their commonality a possible explanation for psychokinetics. He further stated that the psychokinetics could arise if the mental processes of one or more people were focused on meanings that were in harmony with those meanings guiding the basic processes of a particular system.

Other renowned physicists have supported the conclusion that consciousness is a central aspect of quantum physics. John Wheeler has long been an active proponent of observer-created reality. Werner Heisenberg proposed in 1962 that a quantum entity's unobserved attributes are not fully real, but rather exist in an alternative state of being, called potential, until the act of observation promotes some lucky attribute to full reality status. In 1939, Niels Bohr wrote that consciousness must be a part of nature but may have laws that are apart from those of physics and chemistry.

As a consequence, the laws of quantum mechanics alone are not capable of explaining life. To consider biological organisms as machines operating solely by rearrangement of molecules, resulting from local forces, is a serious error. This error becomes even more apparent when we consider our discovery that consciousness connects and affects all things.

The teleological[1] quality of behavior becomes impossible to deny when it is consciously pursued, because we know from direct experience that we often do have a preconceived image of a desired end state toward which we strive. When we enter the realm of conscious experience, we again cross a threshold of organizational complexity that introduces its own new concepts, memories, thoughts, feelings, fears, volitions, and plans. The major difficulty with these concepts is the comprehension of how these mental events are consistent with the laws and principles of the physical universe.

A Nobel Prize winner in neurophysiology, R.W. Sperry in 1966 regarded mental events as holistic configurational properties "that have yet to be discovered, but will be different from, and more than the neutral events of which they are composed...they are emergents of these events." He conceptualized that higher-level entities possessed laws and principles that could not be reduced to lower-level laws. He commented, "These large-cerebral events as entities have their own dynamics and associated properties that causally determine

1. Referring to the basic purpose and design occurring in natural phenomena.

their own interactions. These top level systems have properties that supersede those of the various subsystems they embody."

Thoughts are ascribed definite causal potency, for they can make things happen. Sperry has indicated that the set of laws in the upper-level systems control neural patterns (holistic configurational properties), while the lower set control atoms, which constitute neurons. He felt that mental forces exerted regulatory control in brain physiology. Sperry said that his downward causality did not violate the lower-level laws, and he stated, "The way in which mental phenomena are conceived to control the brain's physiology can be understood very simply in terms of the chain of command of the brain's hierarchy of causal controls. It is easy to see that the forces operating at subatomic and subnuclear levels within brain cells are molecule-bound and are superseded by the encompassing configurational properties of the brain molecules, in which the subatomic elements are embedded."

Sperry's (1966) position holds that the agency of causation can operate simultaneously at different levels and between different levels of complexity without conflict. He goes on to state:

> It is theorized that conscious phenomena are emergent functional properties of brain processing, which exert an active and controlling role as causal determinants in shaping the flow patterns of mental excitation. Once generated from neural events the higher order mental patterns and programs have their own subjective qualities and progress and interact by their own causal laws and principles, which are different from and cannot be reduced to those of neurophysiology. The mental forces do not violate, disturb, or intervene in neuronal activities, but they do supervene, in multilevel and interlevel causation in addition to the one level sequential causation traditionally dealt with.

At some point in this informational system, the flow of information ceases to be tied to local causality but, possessing quantum qualities, becomes nonlocal in causality and nature. The latter informational causality becomes distinctively different from physical causality but remains totally interdependent and complementary—the one process being contained in the other. Expressing it in Bohm's language, the physical (local) causality represents the explicit order, while the quantum (nonlocal) causality represents the enfolded, hidden implicated order.

The physicist Freeman Dyson wrote in 1979, "I think our consciousness is not just a positive epiphenomenon carried along by the chemical events in our brains but is an active agent forcing the molecular complexes to make choices between one quantum state and another. In other words, the mind is already inherent in every election (choice)." Again, since quantum mechanics expands the concept of an unbroken whole to the entire universe, we can now consider that a form of consciousness is attached to everything.

In concluding this section, we are going to return to some further and enormously important considerations offered by David Bohm. He was quite unusual in questioning the primary mechanism for all scientific inquiry. He stressed that thought creates structures and that thought then pretends that they are objective realities, independent of thought. Our "objective reality" is largely a construct of thought; the failure to recognize this leads to an endless circle of self-deception, both in science and in life. As a consequence, Bohm felt that personal and collective suffering has its roots in this kind of human *misthought*.

After analyzing "thought" over many years, Bohm further stated in 1982,

> Thought is really a very tiny little thing. But thought forms a world of its own in which it is everything...It reifies itself and imagines there is nothing else but what it thinks about itself and what it thinks. Therefore, thought will now take the words, "the nonmanifest" and form the idea of the nonmanifest; and therefore thought thinks the manifest plus the nonmanifest will together make up the whole and that this whole thought is now a step beyond thought, you see. But in fact, it isn't. This nonmanifest (that thought imagines) is still the manifest, by definition, because to imagine is also a form of thought.

Bohm had always been willing to seek additional dimensions; in his search for scientific truth, he was led to conduct an extended dialogue with the Indian sage and mystic J. Krishnamurti. As a result of this relationship, Bohm decided that it was vital to go beyond thought and considered meditation as a path of further inquiry. Bohm commented in 1985, "Meditation could even bring us out of all (the difficulties) we've been talking about...Somewhere we've got to leave thought behind and come to this emptiness of manifest thought altogether...In other words, meditation actually transforms the mind...it transforms consciousness." Bohm believed that meditation could put one in contact with that which is beyond thought, which is "enfolded truth."

He came to regard matter, energy, and meaning as the three major constituents of our existence. As Bohm's work matured, he placed increasing emphasis on truth as the essence of meaning!

Summary

In the wake of the Clauser and Aspect et al. experiments, there has been a cascade of advancing quantum theory that has been in near-avalanche mode. David Bohm, who will probably be regarded as the most enlightened quantum physicist of the twentieth century, believed that there were two levels of consciousness: a personal, observed one that is separate, deterministic, and wholly dependent upon the past; and an enfolded, deeper one that is unified, timeless, and ever expanding. The underlying consciousness of mankind is one, and this consciousness is connected to everything in the universe.

He concluded that this greater transformation of the mind could only be brought about by putting oneself in contact with that which is beyond thought: truth. It is cause for celebration to see that a dramatically different and truer vision of reality can escape our stubborn and distorting projective process and become observable. This collective vision has allowed Professor Zeilinger to present us with his most remarkable consciousness-expanding light particle teleportation experiment. This monumental scientific breakthrough has driven the final nail into the coffin of what had been construed to be commonsense reality.

Questions and Answers

Q: What are the central psychological issues that are blocking us from accepting quantum concepts to advance more readily in our quest for a higher level of consciousness?

A: Currently, approximately 85 percent of the world's population has consciousness levels that mainly consist of shame, guilt, apathy, grief, fear, desire, anger, and pride. At these levels, the connected and maladaptive emotions are humiliation, blame, despair, regret, anxiety, craving, hate, and scorn.

These emotions are then expressed as the processes of destruction, abdication, despondency, withdrawal, enslavement, aggression, and inflation (Hawkins 1995). This aggregate at higher levels is characterized by neutrality, willingness, acceptance, reason, love, joy, peace, and eventually enlightenment. These ascendant levels have been further identified and confirmed by Hawkins as the processes of empowerment, release, intention, transcendence, abstraction, revelation, transfiguration, illumination, and finally pure consciousness. The lower levels of consciousness are tenaciously held in place by the narcissistic, murderous, and megalomaniac ego's intent to maintain itself in opposition to the Divine. The ego blocks mindfulness by the unconscious use of denial and repression, while attempting to seduce and reduce awareness by offering specialness, the thrill of being right, and its selfish claim that all thoughts are "mine."

Q: What are the mental and intellectual obstacles that oppose our movement to higher consciousness?

A: The observed world and its scientific corollary, Newtonian physics, have provided locality, causality, and time measurement—a reductionistic framework that is helpful and comfortable but generates and supports a total absence of meaning. Conceptually stated, this process is repetitive and circular, placing narrow limits on any attempts to access higher consciousness levels.

David Hawkins has insightfully recognized the following fatal flaws in our current thought systems that have so limited our progress in consciousness: (1) failure to differentiate between subjective and objective; (2) disregard of the limitations of context inherent in basic design and terminology; (3) ignorance of the nature of consciousness itself; and (4) the misunderstanding of causality in the universe. The consequences of these shortcomings are very apparent when we observe a planet that is obsessively focusing energy on correcting effects rather that identifying cause. As was stated earlier, there are no causes in the observed world; the observed world is a display of effect only.

9
Mysticism

A Transcending Pathway

We have only to believe. And the more threatening and irreducible reality appears, the more firmly and desperately must we believe. Then, little by little, we shall see the universal horror unbend, and then smile upon us, and then take us in its more than human arms.
—Pierre Teilhard de Chardin

Mysticism has been described as the practice of putting the self in direct relation with the supreme power of the universe, the unifying principle of life, by illumination or direct intuition. The nature of mysticism made firsthand objective studies essentially impossible, confining study to biographical and autobiographical accounts. This situation prevailed until the mid-twentieth century.

Religion and mysticism have been linked together rather inseparably. The terms *mysticism* and *mystic* have often been misinterpreted to mean magical or occult. Religious practice in India has frequently been regarded as mysticism, when much of the practice represents an external quiet attitude. Common misleading opinions about mysticism and mystics involve ideas of impracticality, fortune telling, and revolutionary trends. However, careful scrutiny reveals that most mystics possess traits that are quite the opposite. The concept of the mystic as a lonely thinker is not borne out by studies. No generally accepted explanation of mysticism existed prior to recent inquiry.

In the twentieth century, biologist, philosopher, and psychologist Ken Wilber was the first to provide any extensive psychological study of this most fascinating and valuable process. Studies of mysticism have identified two primary representations. The first is to regard the soul as outside of God, arising to God in sequential stages. The other concept is the view that God is within the soul and can be reached by going deeper and deeper into one's inner reality (transcendence). The pathway of transcendence is the most commonly held idea by mystics and forms the nucleus of both ancient and contemporary mysticism. The process of transcendence is in opposition to acquisition of knowledge via sensory perception. Mysticism is both a paradoxical and creative _expression, for it begins as an intense search for a supernormal or metaphysical reality and ends by going far beyond one's individuality to an integration of the self, ending in a sense of pervasive unity and oneness.

Mysticism has been a part of ancient Chinese Taoism, Indian Hinduism, Neoplatonism, and Mohammedan Sufism. In these religions, philosophy and mysticism were often connected. The Middle Ages saw mysticism associated with contemplative religious orders, and prominent mystics were represented in both Eastern and Western churches. Some of the more visible were St. Theresa of Avila, St. Catherine of Sienna, St. Francis of Assisi, Meister Eckert (Johannes Eckert), and St. Bernard of Clairvoux.

Seventeenth century England found mysticism in the writings of Cambridge Platonists St. Thomas More and Ralph Cudworth. Jonathan Edwards, an English Congregational minister, mixed strong mystical elements with religious revivals. Christian and non-Christian mysticism expanded during the twentieth century. The German theologian Baron Friedrich von Hagel (1852-1925), poet and writer Evelyn Underhill (1875-1941), and writer and theologian Ralph Inge (1860-1954) were associated with the Christian tradition. During this period non-Christian mystics included the Buddhist Daisetz Suzuki (1870-1957), the Indian Hindu philosopher Radhakrishnaan (1888-1954), and the Islamic mystic, British scholar Reynold Alleyne Nicholson (1868-1965).

An overview of the mystical systems indicates that Hindu philosophy incorporates the most complete and rigorous discipline yet designed to transcend personal identity, where the goal is the experience of union with the supreme self. In the metaphysical system known as the Vedanta, the individual self (Atman) is joined with the supreme self of the universe (Brahma). The apparent separateness and individuality of beings and events are held to be only a mixture of thought and feelings—in other words, an illusion (maya). This illusion can be collapsed through the realization of the essential oneness of Atman and Brahma. When the religious initiate has overcome meaningless ignorance (avidya), which consists of subject and object, self and no self, the final mystical state of liberation (moksha) is thus attained.

The remainder of this chapter will be a discussion of mysticism and its relationship to physics and the theory of physics. The primary messages of quantum physics are nonlocality and wholeness. The primary messages of mysticism are the nonlocal nature of reality and its ultimate embodiment in wholeness, representing all that is. As a consequence, physics and mysticism have been referred to as "fraternal twins." David Bohm (1980) stated:

> Relativity and, even more important, quantum mechanics have strongly suggested (though not proved) that the world cannot be analyzed into separate and independently existing parts. Moreover, each part somehow involves all the others: contains them or enfolds them...This fact suggests that the sphere of ordinary life and the sphere of mystical experience have a certain shared order and that this order will allow a fruitful relationship between them.

Mysticism is an expression of an inner wisdom reached through an altered state of consciousness. This altered state is analogous to Bohm's enfolded, or implicate, order of physics in that it aligns one with the wholeness of reality. Esoteric mystical traditions are present in all the major religions of the world. Elements of mysticism can be found even in the primitive religions.

While the established religions are obviously quite different in their superficial dogmas, Aldous Huxley found a common core in all of their theologies. He gave this core the name the *perennial philosophy,* and defined it as the transcendental essence of all the main religions perpetuated through this mystical tradition.

The metaphysics of the perennial philosophy, Huxley says, is "immemorial and universal." While the perennial philosophy is sometimes equated with the mystical experience from where it springs, each tradition has its own special insights and beliefs, and each individual's encounter with the unknowable is unique!

Erwin Schrödinger, the Austrian quantum physicist, had a strong interest in mysticism and made the following comment on Huxley's perennial philosophy:

> It is an anthology from the mystics of the most various peoples. Open it where you will and you will find many beautiful utterances of a similar kind. You are struck by the miraculous agreement between humans of different races and different religions, knowing nothing about each other's existence, separated by centuries and millennia, and by the greatest distances that there are on our globe (Margenau 1987).

The mystic who has a direct experience of the Divine ground, representing a fusion between spirit and matter, cannot have that experience validated or invalidated by any of the sciences, or for that matter by any traditional religion. As soon as he or she describes the experience, the oral or written description of the experience is a metaphor or a map of the territory. There is a certain difficulty that arises when one attempts to communicate or compare these experiences. This difficulty lies in the fact that the mystic is confined to a language connected to a level of reality different from that of usual experience. Huxley states this very well when he notes that a truly correct expression of the perennial philosophy is not possible. He said, "Nobody has invented a spiritual calculus in terms of which we may talk coherently about the Divine ground and of the world, conceived as its manifested icon."

In any regard, the body of mystical traditions contains remarkably similar descriptions of a transcendental unity. In all the religions, the perennial philosophy describes an absolute ground, which is the reality of all things. The Absolute is *not* set apart as some sort of creator separated from that which is created. Rather, it is completely whole and indivisible. There is a tendency in the West to identify this wholeness with the philosophy that all things reflect God, but this is incorrect. The Absolute is not just *in* all things. All things *are absolute*. Our usual conception of the universe is a separate manifestation of reality rather than the true reality, which is ultimately indefinable, undifferentiated, and seamless.

Huxley further stated that the basic premise of the perennial philosophy considers the external self as one with the Absolute, and that each individual is on a journey to discover that fact. Each individual's path throughout life is to fulfill his or her destiny and thereby to return to his or her true home. This is accomplished through remembering our true nature rather than through learning. This illustrates Plato's principle of reminiscence, in which he states that prior to birth we had all knowledge; the trauma of birth caused us to forget and then we spend the remainder of our lives seeking this knowledge as *truth*.

In ending this chapter, I would like to point out that an expression of the foregoing exists in the New Testament statement, "The Kingdom of God exists within."

Summary

Mysticism is a spiritual discipline aiming at union with the Divine through deep meditation or trance-like communion. Mysticism teaches that there is a reality beyond perception or intellectual comprehension, but central to being and directly accessible by intuition. Meditation is thought by many to be the royal road to higher consciousness, the consciousness of being and oneness with God.

Most of us think of the mystical experience as a state of mind available only to those who meditate for years, live in some exotic culture, or take psychedelic drugs. The reality, however, is that mystical experiences are more commonplace than we think. They usually occur in fleeting, spontaneous moments, taking the form of a momentary shift in awareness, a sense of déjà vu, or a feeling of connectedness to something greater than the self. These experiences often occur during times of overwhelming emotional intensity, and are especially prevalent during profound interpersonal events like childbirth or falling in love. Higher consciousness is most often experienced not as a bolt from the heavens, but rather as a subtly heightened awareness of your own existence and your place in the world, the universe, and beyond.

Questions and Answers

Q: How significant do you regard the concept of the perennial philosophy?

A: An examination of the recurrent beliefs that occur in the concept of the perennial philosophy reveals a striking and invaluable concordance in regard to the fusion of matter and spirit. These elevated collective concepts truly provide a map of the territory, or a template of advancing consciousness that elicits a definitive, resonating affirmation. It has been pointed out that any authentic theory of reality and experience of Divine ground must enfold the essence of perennial philosophy as well as a progressive order of being: (1) matter; (2) biology; (3) emotional responsiveness and a capacity for abstract thought and collective awareness; (4) unconditional love and bliss; (5) nonduality; and (6) pure awareness, beyond consciousness (Wilber 1982).

Q: What recommendations would you have for someone who is an aspirant on the pathway to expanded consciousness?

A: The goal of any endeavor to acquire an ascendant consciousness must focus on the identification and acquisition of truth. In this process it is necessary to gradually become aware of strong positive attraction fields to the truth and strong negative attraction fields to the untrue.

As a matter of orientation, the negative attraction fields manifest shame, guilt, apathy, grief, fear, desire, and anger, while the positive attraction fields manifest neutrality, acceptance, reason, love, joy, peace, and enlightenment. As one moves his alignment from the negative fields to positive fields, judgment of duality is replaced by the expanding essence of acceptance and love. A required mechanism for affecting this gradual transcendence is the exercise of forgiveness. Forgiveness that acts at the level of unfolding nonduality reflects a knowing that sin in truth is an error. The appropriate response is correction. Forgiveness on a worldly level is the equivalent of love; as a corollary, "love can hold no grievances."[2] A review of the perennial philosophies and the most historically transcendent pathways counsel an ongoing prayerful, reflective, and meditative process. The ultimate goal of this latter process is to provide points of departure from our lower energy fields to our higher energy fields by utilizing altered perception. Completely altered perception can then provide a platform that can facilitate a leap in consciousness.

---

2. *A Course in Miracles.*

10

Science and Mysticism

Unexpected Partners, Unexpected Bridges

We often think that when we have completed our study of "one" we know all about "two" because "two" is one and one. We forget that we still have to make a study of "and."

—Sir Arthur Eddington

Robert Muller—known as "the UN's prophet of hope" and a United Nations director under three secretaries general, including U Thant and Dag Hammarskjöld—was particularly noted for his extraordinary global vision. In his book *New Genesis* (1984), he wrote:

> The scientists have now come to the end of their wisdom. Humans are simply incapable of grasping the vastness of creation and all its mysteries. We cannot understand the notions of the infinitely large, of the infinitely small and of eternity. Even the notions of matter and energy, of objectivity and subjectivity are being challenged today. Beyond the elation at our discoveries, there is a certain despair at our inca-

pacity to comprehend the totality. This is where spirituality… comes in. Science in my view is part of the spiritual process: it is a transcendence and elevation of the human race into an ever vaster knowledge and consciousness of the universe and of its unfathomable, divine character.

A particularly insightful formulation regarding the emergence of physics and mysticism was made by the theologian Lawrence Beynam (1978), who wrote:

Science has experienced perhaps the greatest shift of its kind to date. It is for the first time that we have stumbled upon a comprehensive model for mystical experiences, which has the additional advantage of deriving from the forefront of contemporary physics.

Ken Wilber (1982), in his book *The Holographic Paradigm,* made this observation:

Here, rather suddenly, in the 1970's were some very respected, very skilled researchers…physicists, biologists, physiologists, neurosurgeons, and these scientists were not talking with religion, they were simply talking religion, and more extraordinarily, they were doing so in an attempt to explain the hard data of science itself. The very facts of science…the actual data (from physics to physiology) seemed to make sense only if we assume some sort of implicit or unifying or transcendental ground underlying the explicit data…Moreover, and here was the shock, this transcendental ground, whose very existence seemed necessitated by experimental, scientific data, seemed to be identical, at least in description, to the timeless and space less ground of being, (or "Godhead") so universally described by the world's great mystics and sages, Hindu, Buddhist, Christian, Taoist.

Rupert Sheldrake (1995), botanist and scientist, in his book *Seven Experiments that Could Change the World,* made the following comment in a dialogue with David Bohm:

The interesting thing about the Big Bang theory is that the minute you have to address the question of the origins of the

> laws of nature, you're forced to recognize the philosophical assumptions underlying any sort of science. People who think of themselves as hard-nosed mechanists or pragmatists regard metaphysics as a waste of time, a useless speculative activity...But you can force them to realize that their view of the laws of nature as being timeless, which is implicit in everything they say or think or do, it is in fact a metaphysical view.

It is not surprising that certain contemporary sages and mystics have provided spiritual metaphysical views that have resonated with ideas held by modern science. In some instances they actively engaged in dialogues with the scientists in their search for reality and truth. This group of mystics and sages included the Dalai Lama of Tibet, Lama Govinda, Father Bede Griffiths, and J. Krishnamurti. Similar to quantum physics, metaphysics has stated that underneath observed reality is another order of reality that is unifying, undivided, peaceful, and eternal.

On the physicist's side of the fence, it is of interest to note again that many of this century's most renowned physicists have had strong mystical inclinations. In 1986, Renée Weber, a distinguished professor of philosophy at Rutgers and noted commentator on physics, said:

> A parallel principle drives both science and mysticism. The assumption is that unity lies at the heart of our world and that it can be experienced by man. I believe that this one similarity is so powerful that it transcends the many differences which divide science and mysticism...Such terms as "eloquence" and "beauty" recur regularly in the writings of philosophical scientists like Einstein, Heisenberg, Eddington, Jeans, Schrödinger, Bohr, Feynman, Wald, Bohm, Prigogine, Hawking, Sheldrake and others. Behind the aesthetic demand, I believe there lays a spiritual path. My hypothesis is that it does.

A recent Nobel Prize laureate in physics, Paul Davies, was moved to write in his book *The Mind of God* (1992) that "physics provides a more certain pathway to God than organized religion."

Historically, metaphysics and organized religion have taken a stance toward science that has ranged from distant and cool, to openly hostile, depending upon its perceived degree of threat. There have been scientists who have been the focus of major religious censure, such as Galileo, Copernicus, Newton, Darwin, Freud, and Einstein. Many scientists have returned the favor by maintaining an agnostic indifference. We are starting to see hostilities and counter-hostilities between these two camps being replaced by dialogue, even though the dialogues are taking place between the more unconventional scientists and religious leaders.

Renée Weber has been a visible and important pioneer in examining and facilitating the dialogue between philosophy, mysticism, and science as they have addressed their singular and mutual search for unity. In 1986, she published a powerfully influential and valuable book on this subject, entitled *Dialogues with Scientists and Sages: The Search for Unity*. Science has searched for a grand unitary theory for everything, while mysticism represents the experience of oneness with reality. In 1986, Weber expressed the essence of these two entities, science and mysticism, as two approaches to nature.

While science has strived to explain the mystery of matter and being, mysticism has strived to experience it. They both participate in the search for reality because they both look for the basic truth about matter, as well as the source of matter. The quest for source has been crucial for mysticism, but has long been disregarded by scientific officialdom. During the last fifty to sixty years, a growing number of scientists, particularly those in physics, have begun to address this question. The push for oneness and unity has become the most important central link between the aims of science and the aims of mysticism!

Finding equations that represent coherent laws has been the goal of many scientists, but this has not satisfied the greatest of scientists. For those rare scientists, equations have served only to direct them to the hidden reality that the mathematics expresses. Weber commented in 1986,

> Equations for people like Kepler, Galileo, Newton, Schrödinger, de Broglie, Planck, Einstein, Eddington, Jeans, Heisenberg, Bohm, and others appear to be something of a code word, a disguise for their desire to display the source behind the equations...Pythagoras may have had that in mind when he claimed "God geometrizes," and Galileo, when he said that "God's book of nature is written in the alphabet of mathematics."

These scientists have taken a reverential attitude toward science, deemphasizing its ruthlessness and emphasizing its harmony and beauty.

It has been mentioned that metaphysical mysticism may be more scientific than science in its pursuit of the grand unitary theory, because it strives for a more comprehensive unification. Mysticism has included the questioner in posing the question, but science has not done so consistently. And now quantum physics has greatly facilitated understanding this unification by associating the observer with the observed and making it more difficult to separate them.

Many scientists have only paid lip service to this position, because many in the community of scientists (despite the findings of quantum mechanics) believe they can stand apart from that which they work on. The physicist has been responsible for releasing the immense binding power that has held the atom together by employing his persistence, creativity, and intelligence, but has often deleted any impact upon the self, which then remains unchanged!

An assessment of the many quantum physicists reveals that only two, David Bohm and Rupert Sheldrake, have openly explored the metaphysical as an extension of their search for unity! Weber has pointed out that the mystic has employed the very self in the process of deconstructing and reconstructing reality.

For most of us, the energy needed for the mystic's high-energy consciousness and cognition of inner truth is bound up and not available. The mystic and sage have recognized that energy must be focused and channeled, and as a consequence it no longer fractures and exhausts the self in attempts to hold together the rigid, separate self. This process releases the ego and its energy, opening a channel to the unliberated energy of the universe. In order to accomplish this task, the mystic must possess a very sound and integrated personality.

Both physics and mysticism have forms that are quite different. Physics has occupied itself principally with theory development. Most mainstream physicists see the theory as complete when all the mathematics summates consistently to describe a phenomenon and allow one to make observable predictions. For physicists and scientists like David Bohm, the theory is complete or comprehensive only if it unifies the observer with the observed, subtle matter with dense matter, and these with their source.

For the mystic this form is not part of his orientation, because it imposes limits where his view is that of the unbound. In the matter of content, the aim is similar, for they both attempt to explore the ultimate origin of existence and being. Where science has not been able to provide an answer, mysticism has indicated a direction! The mystical experience proposes that matter has its origin in consciousness. Here, subtle matter has generated and controlled dense matter, forming a continuum. Matter becomes more subtle (moves toward the energy, or wave, state) the closer it moves toward higher consciousness. At their most inward and subtle, matter, energy, and consciousness coalesce and become indivisible.

In the early 1990s, Eugene d'Aquili (a psychiatrist), Andrew Newberg (a radiologist at the University of Pennsylvania), and a clinical associate, Vince Rause, used an imaging technology called SPECT scanning to map the brains of Franciscan nuns in deep contemplative prayer and Tibetan Buddhists meditating. Their scans photographed blood flow—charting levels of neural activity in each subject's brain at the time each person reached an intense spiritual peak. When they studied the scans, the researchers noted a marked alteration in blood flow in an identical portion of each person's left parietal lobe. This area of the brain is responsible for drawing a distinction between self and the rest of existence. At the peak moments of prayer and meditation, blood flow to this parietal area was sharply reduced. The orientation area was deprived of information required to draw a line between self and non-self, or the outside world. The researchers believed that the subject would experience a sense of limitless awareness, merging into infinite space. It appears that they have captured photographs of the brain nearing a state of mystical transcendence, described by all major religions as one of the most profound spiritual experiences. Catholic saints have called it mystical union with God; Buddhists call it interconnectedness. Rause asked this question: "Does this mean that God is just a perception generated by the brain or has the brain been wired to experience the reality of God?" Newberg replied, "The best and most rational answer I can give to both questions is yes." This is why Newberg states that you can't simply think God out of existence. (*Why God Won't Go Away*, Newberg, 2000). A passage from *A Course in Miracles* (1975), a leading source of metaphysical-mystical thought over the last quarter of a century, echoes a similar concept: "The word inevitable is fearful to the ego but joyous to the spirit. God is inevitable and you cannot avoid Him any more than He can avoid you."

In her book *Dialogues with Scientists and Sages*, Weber emphasized the greater completeness and wholeness of mysticism and cited the mystic's altered state of consciousness, which has aligned and harmonized his awareness with nonmaterial energy and its intelligence. The mystic is in time with the deeper structure of nature, where transformation and flux are ongoing! Using this criterion, "the search for unity," she speculates that since the mystic as observer merges more completely with the observed and aims at a more comprehensive unification of space, matter, and consciousness, mysticism may be considered more scientific than science.

In the Buddhist and Tibetan practice of mysticism, the meditator visualizes a beam of light the size of a hair within the person's center, which is then allowed to extend to the whole body and then again to all outer space—extending as energy and light to all of the cosmos. Here, the differentiation between inner and outer space, self and nature, matter and consciousness, has lost all boundaries. Weber states, "The scientist makes the dense matter dance to produce pure energy; the mystic, master of subtle matter, dances the dance himself."

Summary

In spite of Professor Renée Weber's emphasis on the greater whole of mysticism, the immense power and value of science remains. There is considerable justification for viewing science and mysticism as unique and irreplaceable perspectives on reality. Science and mysticism have grown closer and closer in the quantum and post-quantum eras. The way science has begun to regard time in its search to comprehend the Big Bang is to push time back to the beginning of time and the timeless, placing itself in the courtyard of mysticism.

Questions and Answers

Q: It has been stated that science has searched for the mystery of matter and being while mysticism has strived to experience it. What are some of your personal observations concerning this seeming duality?

A: The impetus for the resolution of a scientific approach to consciousness versus a metaphysical, mystical approach is aided by the compelling power that both disciplines are seeking unity and oneness. Historically, science has had its origin in the locality, linearity, and causality of measurement. At this level, there is a psychic overlay of investment connected to the predictive value of measurement and its familiarity. This psychic investment wishes to propagate itself and is threatened by fear of extinction when faced with the veritable fact that nonlinearity, nonlocality, and noncausality are a much truer reality. Some scientific observers have reacted by stating, "I don't care if it's true, I don't believe it." A more sanguine, less fearful, and more aware attitude would be to regard Cartesian and Newtonian positions as the connected lower rung on the ladder of consciousness. This positionality can provide a more embracing context, serenely allowing awareness to expand—circumventing a sense of loss and fear while becoming congruent with an unfolding joining and unifying field of strong, uplifting attractors.

Q: Rupert Sheldrake and David Bohm have fused mysticism with their quantum physics research; in accomplishing this, both—but particularly Bohm— have extensively used a dialogue process. How would you assess the dialogue aspect of this process?

A: The dialogue process is central to the participatory interactive continuum. This continuum provides an intentional reoccurring recontexualization that securely and peacefully embraces the desire and quest for expanding knowing. In this paradigm, the harmonious exchange of observer/observed positionalities allows inquiry to leave the limiting field of dividing informational objectivity and to enter the nonlocal, nonlinear, nontemporal, unlimited field of consciousness-expanding subjectivity. Here, pure awareness waits, enfolded within the Divine ground of the Absolute, its Source.

11

Science and Mysticism

The Ladder of Consciousness

He who lives in the present lives in eternity.
—Ludwig Wittgenstein

In order to further develop the main concepts presented in the last chapter, we will concentrate on the integration of science and mysticism by reintroducing you to Ken Wilber, who has addressed the attainment of higher consciousness through the integration of science and spirituality. In so doing, he has displayed a most unique creativity and lucidity. He has distinguished himself as a transpersonal psychologist who has successfully extended his studies into other areas; namely, philosophy, mysticism, and particularly quantum physics. Wilber has defined transpersonal psychology as "a sustained and experimental inquiry into spiritual, transcendental, or perennial philosophical concerns." As a consequence, he has presented transpersonal psychology as the contemporary expression of the perennial philosophy that we discussed in the previous chapter. Wilber has discerned that all formulations of the perennial philosophy provide for a hierarchical chain of consciousness in which the spirit-self journeys from the lowest, most dense and fragmentary level to the highest level, which is the source and nature of all the levels.

Wilber has used the term *involution* for the movement of the highest order of consciousness down through the consciousness spectrum to create the manifest, or material, world; this is the process by which the higher levels of being are involved in the lower levels. Interpreting the perennial philosophy, he uses the term *microgeny* to represent the unfolding and enfolding, moment-to-moment movement of the spectrum of consciousness.

At this point, we will focus on the shared features of this hierarchical chain of consciousness, particularly the lower levels, where they find a common ground with physics. The basic feature of the perennial philosophy is that consciousness is displayed as a hierarchy of dimensional levels. The number of levels is not significant; the interesting point is that the lowest level is only an aspect of, and is subordinate to, the level directly above. Bohm has frequently commented that consciousness is made up of a continuum of *ordering* principles. The simple version is the lowest grade and is equated with the Newtonian outlook. The lowest level is exactly what we know through our five senses; as consciousness progresses upward, it ascends to the level of the mystical.

As was stated earlier, the lowest level of consciousness is the densest; it's material-bearing and fragmentary, and is separated by the greatest distance from ultimate consciousness. Wilber has labeled this level *insentient consciousness*. The higher we go, the less dense and more holistic his levels become. Density is used here in a manner similar to, and inversely related to, Bohm's concept of subtlety. Density is the expression of the unfolded, explicate order, while the subtle and nonmaterial is equaled with enfolded, implicate order.

It is essential to keep in mind that levels of consciousness are not separated and discrete, but rather are mutually interpenetrating, interconnected, and interdependent. In 1982, Wilber quoted A.P. Shepherd (1977):

> These "worlds" are dimensional levels and are not separate regions, specially divided from one another, so that it would be necessary to move in space in order to pass from one another. The highest worlds completely interpenetrate the lower worlds, which are fashioned and sustained by their activities.

What divides or specifies the levels of consciousness is *focus*. The lower level represents a more limited focus than the levels above. Because of this limitation, the consciousness on the lower planes cannot experience and is not aware of the worlds above it. This remains true even though the higher worlds interpenetrate and sustain the levels below. On the other hand, a consciousness can move up to a higher level by broadening its focus. Then the higher world becomes manifest, and the consciousness exists on a new plane.

Although all levels of consciousness are available with the proper vision to perceive them, individuals tend to create boundaries of existence that limit them to focusing on one level only or on those levels below it. Huston Smith uses a metaphor in his book *Beyond the Post-Modern Mind* (1989) that is applicable here. He says that the divisions between the stages of consciousness are like one-way mirrors. If we look up, we see only the reflection of the level we now occupy. If we look down, the mirrors are as transparent as plate glass. When the self reaches the highest plane, even the glass is removed, and pure interpenetration exists. If, as Huxley says, the eternal self is on a journey, that journey can be seen as a process that widens the boundaries of comprehension.

The hierarchic chain of consciousness has different numbers of levels, depending on the version of the perennial philosophy. There may be three levels (matter, mind, and spirit), or dozens of levels. The number is less important than the fact that consciousness is displayed as hierarchy. Wilber describes six levels of evolved consciousness:

1. Ultimate: consciousness as such, the source and nature of all other levels
2. Causal: formless radiance, perfect transcendence
3. Subtle: archetypal, trans-individual, intuitive
4. Mental: ego, logic, thinking
5. Biological: living (sentient) matter/energy
6. Physical: nonliving matter/energy

We will look at the multilevels of consciousness in greater depth because a deeper comprehension of these concepts will prove to be invaluable for understanding quantum reality. As was stated earlier, each level transcends, but also includes, all lower levels. Because the higher level transcends the lower, the higher cannot be derived from or explained by the lower. And while a higher level contains the attributes of the lower level, it also has new aspects that are clearly different from those of the lower and cannot be seen as a simple derivation of the lower plane. This process is a sharp departure from nineteenth-century science, which wished to reduce elements into smaller and smaller units.

Wilber uses the metaphor of a ladder to describe the hierarchic structure. From each rung, the view of reality is broadened, more inclusive, more unified, and more complex than from the previous rung. To use a dimensional analogy, the three dimensional sphere contains the two-dimensional circle, but not vice versa. This idea is similar to Bohm's concept of the spectrum of order. Climbing Wilber's ladder broadens the context. In Bohm's construct, what seems to be random on one rung, becomes more ordered on a higher rung.

Wilber subdivides each level of consciousness into a *deep structure* and a *surface structure*. The deep structure contains all the potentials of that level along with all its limits. In essence, the deep structure is a paradigm, and as such it contains the whole set of forms for that level and the levels below it. The limiting principles within the deep structure determine which surface structures are actualized. Wilber defines the *ground unconscious* as all the deep structures of all the levels existing as potentials ready to emerge. So in a sense, each deep structure contains the potential of all deep structures in an enfolded order. All the deep structures of all levels might be considered the counterpart to Bohm's holomovement.

The surface structure of any level of consciousness is a particular manifestation of the deep structure. While the surface structure is constrained by the paradigm of its deep structure, the surface structure can manifest, or unfold, any potential within the limits of that deep structure. In *A Sociable God* (1986), Wilber compares the relationship between the surface and deep structures to a game of chess. The surface structures are the various pieces and the moves they make during the course of the game. The deep structures are the rules of the game. The deep structure, via the "rules," holistically unites each piece to all the others. Each game played has a different surface structure, but all games share the same basic deep structure. Thus, we can say that the deep structure is the nonmanifest order; it does not have separation in space or time. At least on the lowest, material level, the surface structure is the explicated, manifest portion of the material world. The deep structure is interpenetrating and interconnected, but the surface structure is relatively separated, with a space-time attribute.

Wilber defines any movement of the surface structure as translation. Again using the chess game as an analogy, translations are moves in the game. As for what causes the moves in the game, Wilber says that they are both (1) the sum of previous moves in the game thus far, and (2) the judgment of the player in reference to previous moves. He indicates that translations are historically conditioned but are not totally determined. There appears to be a mixture of causal and creative components, and somehow the judgment of the player is a factor. Furthermore, a particular level maintains itself by a series of more or less constant translations.

Wilber calls the movement from one deep structure to another *transformation.* What is it that is doing the moving from one level to the other? None other than the self, the player of the game of chess.

The self is defined as a level of consciousness; its attributes are those of the deep structure of that particular level. The self, Wilber says, is involved in what he calls the *Atman project*—Atman (Sanskrit) being the ultimate nature of reality. It is the final unity, the fundamental whole, the source and suchness of reality. It is the top level of consciousness and, at the same time, all levels. It is also indefinable.

The Atman project is the production of ever-higher unities, the ascension of Wilber's ladder of consciousness, encompassing the ladder as the ascent takes place. The goal is to bring all levels into the top level, where there is a realization that there were no levels in the first place. The self is on all levels all along, but remains ignorant of this by a self-inflicted boundary that limits awareness! The ascent is a widening view of the self that eventually becomes all-encompassing, like the mythic snake swallowing its own tail until nothing remains. That nothingness is everything, and all that is.

Assuming that the self is engaged in the Atman project—playing the game of chess, climbing the ladder, or swallowing its tail—how is this done? Does that self lift itself by its bootstraps to the next level? At each level of development, Wilber says, the self maintains itself by a series of constant translations. At each level, an appropriate symbolic form emerges and mediates or assists the emergence (through differentiation) of the next higher level. The self then focuses on or identifies with a newly emergent, more complex, and more unified higher structure, which was there all along. According to Wilber's spectrum of consciousness, the biological stage, which is sentient, emerges and ascends from the physical, or nonsentient, material world.

Wilber describes the process as follows:

> As evolution proceeds...each level in turn is differentiated from the self or "peeled off" so to speak. The self, eventually disidentifies with its present structure so as to identify with the next higher-order emergent structure. More precisely (and this is a very important technical point), we say that the self detaches itself from its exclusive identification with that lower structure. It doesn't throw the structure away, it simply no longer exclusively identifies with it. The point is that because the self is differentiated from the lower structure, it transcends that structure (without obliterating it in any way), and can thus operate on the lower structure using the tools of the newly emergent structure. Thus, when the body ego (the mental ego that merges to create an individual organism) was differentiated from the material environment, it could operate on the environment using the tools of the body itself (such as the muscles).

In effect, Wilber is saying that the self is now on a higher level than the material universe and can use the tools of that higher level to operate in the material universe. The self is now the chess player and can arrange and produce structures through translations of the surface structure. The self is constrained only by the deep structure of the material world and is conditioned by a history of such translations. We should emphasize that although the body is in the material world, and the self can operate in and on the material world, it is identified with the next level up; its activities are now limited by the deep structure of the second level; but when operating within the lowest level, its activities are constrained by the deep structure of the level.

Since it transcends the lower level, the self can both operate in and integrate the lower level. The identification with the next highest level (biological) allows the self for the first time to see that it was identified with the lower level but has now broken away. While it was identified with the material world exclusively, it could not operate in the world, since it did not recognize its identification. An example of this process is found in infant development. Wilber observes that at the end of the sensorimotor period of growth, a child can move objects in a coordinated fashion. However, this is not possible before the child has become differentiated from its physical environment. As long as the child is identified with its objective surroundings, it cannot operate on them. As Wilber states:

> At each level of development, one cannot totally see the seer. No observing structure can completely observe itself observing. One uses the structures of that level as something to perceive and translate that world, but one cannot perceive and translate those structures themselves, not totally. That can occur only from a higher level.

Of the first three levels of consciousness defined by Wilber, the second, or biological, level is the first one to display the attribute of life. It is represented by very simple biological systems such as cells. At this level there are no concepts, logic, or ideas. The human mind occupies the third, or mental, level. If the human mind is to operate on the material level, it must do so by organizing and integrating the biological level, which in turn operates or integrates the material level. Simply stated, if the mind wishes the body to run or walk, it must operate and integrate the simpler biological systems of the brain and legs, which operate and integrate the atoms and molecules of the material world.

Wilber sees the process of psychological development as mediated by symbolic forms. These forms assist the self, through differentiation, to rise to the next highest level. A symbol is defined as that which points to, represents, or is involved with an element of a different level, either higher or lower. Wilber quoted Huston Smith, who remarked, "Symbolism is the science of the relationship between different levels of reality and cannot be precisely understood without reference to each other." Each deep structure has its own symbolic matrix within which translations of surface structures can unfold and operate.

Language is one of the symbols used by the child to differentiate itself from the biological level in order to transcend to the mental level. Language provides concepts that can be used to operate on both the body and the surrounding world. Note that the symbols themselves are not material but reside on the mental level. Because of this, they are more creative, more complex, and more unifying; they are not just representations of the material level. The mind is free to operate on or translate these symbols directly without having to perform the inefficient and cumbersome operations of the material level itself.

Mathematics is an example of a symbolic system at the mind level. It can be used to represent elements on the material level but, more important, it is more creative, more complex, and more unifying than the material-level elements themselves. We have seen that, according to the perennial philosophy, the deep structures of the various levels are one with the ground unconscious, which contains all the deep structures as potentials ready to emerge. This concept may shed some light on the question of why mathematics works in science and whether mathematics is discovered or created. According to Plato, and more recently Gödel, mathematics is discovered. The perennial philosophy would tend to favor such a position since symbols are potentials waiting to emerge. Furthermore, it is not at all surprising that mathematics works in science once the relationship of the levels is identified.

The mind level is capable of producing symbols, and since the mind level is in the symbolic mode, it can form symbols of elements in levels above as well as below. When symbols are used for elements of levels above, the results are eventually paradoxical. At the highest level of consciousness and awareness, symbols are unnecessary because their essence can be directly known without the necessity for transitional representations (symbols).

Another way of looking at this concept is to recognize that symbolic thought is, by its very nature, rational and so requires a subject (knower) and an object (known). Communication of the subtle level being direct—not mediated by symbols or thought—it does not require subject and object; it grasps reality as a whole. Since the symbols of the mental level are inadequate to represent the subtle, the subtle level can never be entirely understood through the mental level's concepts of logic and thought. To reach such understanding requires direct access, insight, or transcendence to the subtle level. Even when some insight to the subtle level is obtained through meditation, intuition, or psychic powers, the knowledge still must be described, represented, or symbolized with tools of the mental level in order for the experience to be communicated. This led Huxley to conclude that an operational calculus for the Divine ground is not possible.

Objects appear to function as symbols of a physically tuned consciousness. As higher and higher stages of consciousness are reached, symbols become less and less necessary, and eventually fade out. This appears to be a particularly unpopulated area, as it has been reported that representations blink on and off and finally disappear. In this stage of consciousness, the soul finds itself alone with its own feelings—stripped of symbolism and representations—and begins to perceive the gigantic reality of its own knowing. Here it experiences directly that which is beyond the imagination of most of us.

As was pointed out earlier, Huxley stated that the basic premise of the perennial philosophy is that the eternal self is one with the Absolute and that each individual is on a journey to discover that fact. This journey correlates with Wilber's Atman project, where Atman is the ultimate reality of nature. Atman is at the top of the ladder of consciousness and at the same time on all the rungs. Each human soul is climbing the ladder to return home, to return to Atman. But since Atman is also the rungs, paradoxically the soul never left home and needs only to become aware of that fact. Thus, the journey is in actuality the unfolding of higher structures of consciousness! These higher structures are not created, but are enfolded in the lower levels.

The following question then arises: How did the higher levels become enfolded in the lower ones? This process of enfolding is called involution, as opposed to the Atman project, or evolution. If involution is the enfolding of higher levels into the lower states, then at the very lowest state, the material level, all of the levels are enfolded as undifferentiated potential.

All this enfolded potential Wilber labels the ground unconscious. It contains all of the deep structures of all of the levels. At the end of evolution, all of the structures enfolded in the ground unconscious will have unfolded, thus draining the ground unconscious and leaving only Atman. Wilber has provided some commentaries about involution, even though the process is essentially beyond description. The following reflects his distillation of a number of mystical traditions that describe the beginnings of involution:

> The essence of this literature, although it seems almost blasphemy to try to reduce it to a few paragraphs, is that "in the beginning" there is only Consciousness as Such, timeless, space less, infinite and eternal. For no reason that can be stated in words, a subtle ripple is generated in this infinite ocean. This ripple could not in itself detract from infinity, for the infinite can embrace any and all entities. But this subtle ripple awakening to itself, forgets the infinite sea of which it

> is just a gesture. The ripple therefore feels set apart from infinity, isolated, separate.

The creation of the ripple begins the process of involution. At this stage the ripple is very rarefied, or to use Bohm's terminology, extremely subtle. According to Wilber, the ripple is now on the causal level and as such is still quite close to the Absolute. Even so, the first inkling of selfhood is established. It is this sense of selfhood that propels the involutionary process. The self is now paradoxically trapped. On the one hand, it wishes to return to the Absolute to restore it to profound peace. But to do so, it literally must die, for it must give up its sense of self, a terrifying prospect. As a result, the self seeks fulfillment by a compromise. As Wilber put it,

> Instead of finding actual Godhead, the ripple pretends itself to be God, cosmocentric, heroic, all-sufficient, immortal. This is not only the beginning of narcissism and the battle of life against death, it is a reduced or restricted version of consciousness, because no longer is the ripple one with the ocean, it is trying itself to be the ocean.

This involution process is just the reverse of Wilber's Atman project. The ripple creates more restricted orders of consciousness. That is, it descends from the causal level of perfect transcendence to the subtle level by reducing the scope of its consciousness. But its desire for infinity is not satisfied at this level. So its scope is again reduced—to the mental level—and goes into "insentient slumber." At this point, Wilber inserts the following reminder:

> Yet behind this Atman project is the ignorant drama of the separate self, there nonetheless lies Atman. All of the tragic drama of the self's desire and mortality was just the play of the Divine, a cosmic sport, a gesture of Self-forgetting so that the shock of Self-realization would be the more delightful. The ripple did forget the Self, to be sure—but it was a ripple of the Self, and remained so throughout the play.

This entire involution process enfolds all the structures in the ground unconscious. The stage is now set for reversing the process, the Atman project. The deep structures now contain all the undifferentiated potential needed to return to Atman. Instead of restricting consciousness, the self must recognize that it is actually one with the Absolute and thereby expand its consciousness through its journey.

The highest level of the spectrum of consciousness is timeless and spaceless. This concept is difficult to grasp since we are producing the mind level. But as we have seen from Bohm's concept of the hologram, the infinite is present at every point of space.

In Wilber's 1983 discussion of this, he says,

> Since the infinite is present in its entirety at every point of space, all of the infinite is fully present right here. In fact, to the eye of the infinite, no such place as there exists (since, put crudely, if you go to some other place over there, you will still only find the very same infinite as here, for there isn't a different one at each place). Similarly, the Absolute is present at every point in time since the Absolute is timeless. Being timeless, all of Eternity is wholly and completely present at every point of time—and thus, all of Eternity is already present right now. To the eye of Eternity, there is no then, either past or future.

Thus, eternity bears the same relationship to time as infinity does to space: All of time is at the present moment, and all of space is at each point in space. This means that each point in space is identical to all other points, since all points contain the infinite. In these terms, space is nonexistent. In a similar manner, time exists only now. Past and future are human constructions.

Another way of looking at this difficult concept is to envision the entire Absolute as being present at every point of space and time. The following example may help: In our three-dimensional universe, an event may occur at point A. If the knowledge of that event is transmitted with the speed of light, it arrives at point B some time later. This information can be transmitted only within a time not less than it takes for light to travel from A to B. This is the limiting velocity in our three-dimensional world. However, if we operated on a level where this velocity restriction was not present, then point B could receive the information from point A instantaneously. In such a situation, every point is connected to every other point without a time restriction. The following corollary is also true: All points in space are contained in each point. As Wilber puts it, "The entire Absolute is completely and wholly present at every point of space and time..."

The concept of time derives from the fact that we have divided reality up into bits and pieces. In so doing, we create a subject and an object with a space in between. We experience these objects, these aspects of reality, in a piecemeal manner and thus create a linear succession in time. The space less infinite removes the space between subject and object: The timeless now removes linear succession. Instead of knowing the universe from a distance, the self, on the higher levels, knows the universe by being it without the need of space and time.

The goal of the Atman project is to ascend the ladder of consciousness and to provide even higher unities while bringing all levels of consciousness to the top level. Ironically, the climb is an illusion. The higher self exists on all levels all the time; all boundaries are self-imposed. We must learn to widen our focus and thus rid ourselves of these boundaries.

Summary

Wilber's concepts of hierarchy, enfoldment of levels, and the infinite nature of the hierarchy all are similar to concepts developed by David Bohm. The notion of the deep and surface structures obviously parallels Bohm's implicate and explicate orders. The terms ground unconscious and holomovement also share common elements. Also, Wilber's concept of microgeny and Bohm's unfolding and enfolding are comparable. Bohm, using the concepts developed in contemporary physics, and Wilber, using an amalgam of mystical traditions, have arrived at amazingly similar ideas.

Questions and Answers

Q: It's been stated that what separates or divides levels of consciousness is focus. What can be done to expand this focus?

A: It's of great value to appreciate as fully as possible that these levels on a deeper basis are completely interpenetrating, interconnected, and interdependent. This appreciation can help remove some of our strong predisposition to regard everything as divided and separate, keeping our focus myopic and frozen. To depart from a narrower focus to a broader one poses threats of sacrifice and loss. However, on an experiential basis, if one cultivates a level shift from antagonistic anger to releasing neutrality, the self integrates at this higher level. At the higher level of awareness, the self enjoys the greater peace of increased nonattachment; earlier satisfactions now present little attraction.

Q: The material in this chapter implies that a strong dedication is necessary to transcend levels of consciousness. Could you comment on this?

A: This evolving journey of awareness does require a dedicated intention within a supporting matrix. This supporting matrix must designate that this journey is not just invaluable but also that everyone is in need of it. This can be a kindly process that will reconcile opposites and correct the errors in perceptions that reside at the lower levels of consciousness. The consistent choice of forgiveness, peace, and love will lead us out of this observed house of mirrors. The commitment necessary for an expanding awareness is one that constantly alters the context of the goals and meaning of one's life to maintain a metaphysical, spiritual focus. Dividing this process from ordinary life is an error to be avoided.

12

Science and Mysticism

Quantum Dialogues

Inspiration does not come like a bolt, nor is it kinetic energy striving, but it comes to us slowly and quietly all the time.

—Brenda Ueland

The previous chapter provides a theoretical framework for integrating science and spirituality as a means of attaining higher consciousness. This chapter will present joint efforts by scientists and mystics to actually accomplish that task.

Rupert Sheldrake and David Bohm are two scientists who have had particularly active dialogues with mystics. Bohm, whom I regard as the most creative and important quantum physicist of the twentieth century, entered into an ongoing dialogue with the spiritual master and mystic J. Krishnamurti—a dialogue that spanned a period of over twenty years, ending only upon Krishnamurti's death in 1986.

During the last half of the twentieth century, both physicists and mystics increasingly used dialogue as an exploratory process, since it is ongoing and open-ended, and it maintains a participatory mode. Dialogue represents the insights of each person at one moment in time and does not disallow other responses later in time. In this context, dialogue is a uniquely creative process. Some physicists, by the nature of their writings, entered into an open dialogue with others, and in certain cases their dialogue has had a very mystical orientation. Those physicists whose written dialogue had a marked mystical component were Werner Heisenberg, Niels Bohr, John Wheeler, and Ilya Prigogine.

Heisenberg used the idea of dialogue as an analogy for his scientific expression, as did John Wheeler with his idea of a "participatory universe." In addition, Ilya Prigogine used the phrase "A New Dialogue with Nature" as a subtitle for one of his books. Renée Weber, professor of philosophy at Rutgers University, has made ongoing efforts to encourage dialogues between scientists and mystics. Her book entitled *Dialogues with Scientists and Sages: The Search for Unity* provides epic trailblazing in this area. She commented:

> Participants in dialogue, besides using their own intelligence and insight seem to draw on some insight beyond ourselves. It is this factor which promotes dialogue to the sacred. For Socrates, dialogue allowed a link with a divine presence within and placed it in the realm of the sacred.

David Bohm has proposed that meaning is a form of being, and when we act to interpret the universe we create a change in nature through our meanings. The process of meaning transforms man into nature's partner, participating in the process of evolution. Bohm further suggested that as we dialogue about the cosmos, our questions, doubts, and truths are forms that lead the cosmos toward further clarity and truth. Through man, the universe questions itself and attempts different answers for itself, paralleling our own process as it attempts to decode itself. It would appear that this is the deeper message of both quantum physics and metaphysical mysticism!

Renée Weber feels that it is probably too much to expect, even of the most visionary of the scientists, to reach the mystic's cosmic oneness, but the process of science can take us within striking distance of ethics. The awareness of unity and interconnectedness with all being leads to an empathy with others and the development of ethical responsibilities. She states:

> It [empathy] expresses itself as reverence for life, compassion, a sense of the brotherhood of suffering humanity, and the commitment to heal our wounded earth and its peoples.

> All the mystics (and virtually all the scientists) in these dialogues draw this connection between their vision of the whole and their sense of responsibility for it. (Weber, 1986)

We will provide several excerpts from dialogues to illustrate the development of an expanded consciousness, with the creative insights that derive from this process. David Bohm, in particular, thought that our culture reinforces our fragmented and broken ways of thinking and that through a free-form dialogue, genuine and creative collective consciousness could be reestablished.

Many feel that Bohm's preeminence is related to his willingness to extend his search for answers far beyond the confines of physics into art, philosophy, and spiritual mysticism. Bohm was willing to leave behind everything he knew in his search for new clues and insights. He demonstrated his commitment to wholeness not only in his theories of physics, but also in his investigation of the nature and origin of knowledge. In the 1960s, Bohm inquired into the nature and function of order in art and engaged in a seven-year-long correspondence with the American artist Charles Biederman. Through their correspondence, Bohm became aware of the hidden and underlying order in post-impressionist painting, conceptualizing it as equivalent to the order in quantum theory, which he had named the implicate order.

The following is a dialogue between the philosopher Renée Weber and physicist David Bohm, entitled, "Mathematics: The Scientist's Magic Crystal" (1986). Here the relationship between science and mysticism is explored, providing further support for the existence of a hidden quantum reality:

Weber: Mathematics is pure thought.

Bohm: That's right. You won't find it anywhere in matter.

Weber: You are saying that even today's physicists, who might be least inclined towards anything spiritual, are practically forced to assume that it is beyond the material.

Bohm: Tacitly, anyway. Physicists may not accept this, but they are attributing qualities to matter that are beyond those usually considered to be material. They are more like spiritual qualities insofar as we say there is this mathematical order which prevails, which has no picture in material terms that we can correlate with it.

Weber: Is it an aesthetic principle or something deeper still that makes them hold out for one rather than for three or four ultimate laws? Is it a spiritual drive, without their realizing it?

Bohm: It is probably a universal human drive, the same one which drives people to mysticism or to religion or art.

Weber: You're saying that this underlying drive, which even the materialistically oriented physicist possesses, cannot be explained in terms of prediction and control.

Bohm: They would find it rather boring to say: "We do nothing but predict and control." If you talk to any of them, Penrose or Hawking, for example, I doubt if they would be satisfied with that.

Weber: But neither would they be happy if we called them unconsciously mystical.

Bohm: No, they would regard that as absurd. But that may be an overlay of language. People in science may be forced to adopt a hard-boiled language to meet their hard-boiled colleagues.

Weber: To be respectable within the peer group.

Bohm: Yes, and gradually they come to believe it themselves.

Weber: So you feel that behind their empirical pursuit of physics they are searching for this underlying unity.

Bohm: Otherwise, why should they be interested at all? Take Stephen Hawking, who is hardly able to do anything due to his illness, yet he is driven to understand what lies behind things. He doesn't merely want to predict and control nature. Why does he drive himself like that?

Weber: He would probably say, "I want to understand."

Bohm: Yes, but you might understand things in terms of plurality: Why drive for unity?

Weber: He might say, "It's a sign that we're getting closer to the way things work." He would use a rational justification, not a spiritual one.

Bohm: He might say that. But I think he and others have an aesthetic appreciation of the unity.

Weber: Feynman said that those who don't understand mathematics don't realize the beauty in the universe. Beauty keeps coming up together with order, simplicity, and other Pythagorean and Platonic categories.

Bohm: Order and simplicity and unity, and something behind all that which we can't describe.

Weber: Do the great minds working in physics sense something of that?

Bohm: Yes...

Weber: Both the scientists and the mystic see in matter as you are using the word, something that is both immanent and transcendent.

Bohm: The mystic sees in matter an immanent principle of unity, and this is implicitly what the scientist is also doing...Matter was found to be far more subtle than was supposed, both for quantum mechanics and relativity.

Weber: Does "subtle" imply spiritual?

Bohm: It moves in that direction.

The most striking example of Bohm's inclination to go deep into other knowledge-based (epistemological) realities was his extensive dialogue with J. Krishnamurti. Both Bohm and Krishnamurti were completely committed to the (observer-observed) continuum that is so crucial to both quantum theory and mysticism. Bohm and Krishnamurti developed a close friendship, and they carried on an intensive dialogue over many years (1965–85) that entailed deep explorations into a variety of ultimate meanings that relate to underlying reality.

The following dialogue excerpts occurred between Bohm and Krishnamurti in 1985. In this first one they explored the function of time and insight:

Bohm: ... You could say time is a theory which everybody adopts for psychological purposes.

Krishnamurti: Yes. That is the common factor; time is the common factor of man. And we are pointing out time is an illusion...

Bohm: Psychological time.

Krishnamurti: Of course, that is understood.

Bohm: Are you saying that when we no longer approach this through time, then the hurt does not continue?

Krishnamurti: It does not continue, it ends¼because you are not becoming anything.

Bohm: In becoming you are always continuing what you are.

Krishnamurti: That's right. Continuing what you modified...

Bohm: That is why you struggle to become.

Krishnamurti: We are talking about insight. That is, insight has no time. Insight is not the product of time, time being memory, etc. so there is insight. That insight being free of time acts upon memory, acts upon thought. That is, insight makes thought rational, but not thought which is based on memory. Then what the devil is that thought?

Krishnamurti: No. Wait a minute. I don't think thought comes in at all. We said insight comes into being when there is no time. Thought, which is based on memory, experience, knowledge...that is the movement of time as becoming. We are talking of psychological and not chronological time. We are saying to be free of time implies insight. Insight, being free of time, has no thought.

The following conversation demonstrates the far-reaching aspects of the Bohm- Krishnamurti dialogue as they inquire into fragmentation, division, and solutions to psychological problems:

Bohm: If we accept that we are fragmented, we will inevitably want to be totally secure, because being fragmented we are always in danger.

Krishnamurti: Is that the root of it? This urge, this demand, this longing to be totally secure in our relationship with everything? To be certain? Of course, there is complete security only in nothingness!

Bohm: It is not the demand for security which is wrong, but the fragmentations. The fragment cannot possibly be secure...

Krishnamurti: Of course.

Bohm: Thoughts he doesn't know. He is not actually free to take this action because of the whole structure of thought that holds him...

Bohm: What is wrong?

Krishnamurti: The way we are living.

Bohm: Many people must see that by now.

Krishnamurti: We have asked whether man has taken a wrong turning, and entered into a valley where there is no escape. That can't be so; that is too depressing, too appalling...

Bohm: Do you perceive in human nature some possibility of a real change?

Krishnamurti: Of course. Otherwise everything would be meaningless; we'd be monkeys, machines. You see, the faculty for radical change is attributed to some outside agency, and therefore we look to that, and get lost in that. If we don't look to anybody, and are completely free from dependence, then solitude is common to all of us. It is not an isolation. It is an obvious fact that when you see all this...the stupidity and unreality of fragmentation and division...you are naturally alone...

Bohm: People feel they want something that really affects us in daily life: They don't just want to get themselves lost in talking, therefore, they say, "All these vapid generalities don't interest us." It is true that what we are discussing must work in daily life, but daily life does not contain the solution to its problems.

Krishnamurti: No. The daily life is the general and the particular.

Bohm: The human problems which arise in daily life cannot be solved there.

Krishnamurti: From the particular, it is necessary to move to the general; from the general to move still deeper, and there perhaps is the purity of what is called compassion, love, and intelligence. But that means giving your mind, your heart, your whole being to this inquiry.

This last excerpt is from a dialogue between Bohm and Krishnamurti on holomovement and death:

Krishnamurti: What is movement, apart from movement from here to there, apart from time...is there any other movement?

Bohm: Yes.

Krishnamurti: There is...Is there a movement which in itself has no division? Would you say it has no end, no beginning?

Bohm: Yes...Can one say that movement has no form?

Krishnamurti: No form, all that. I want to go a little further. What I am asking is, we said that when you have stated there is no division, this means no division in movement.

Bohm: It flows without division, you see.

Krishnamurti: Yes, it is movement in which there is no division; do I capture the significance of that? Do I understand the depth of that statement? I am trying to see if that movement is surrounding man?

Bohm: Yes, enveloping.

Krishnamurti: I want to get at this. I am concerned with mankind, humanity, which is me...I have captured a statement which seems so absolutely true...that there is no division. Which means that there is no action which is divisive?

Bohm: Yes.

Krishnamurti: I see that. And I also ask, is that movement without time, et cetera. It seems that it is the world, you follow.

Bohm: The universe.

Krishnamurti: The universe, the cosmos, the whole.

Bohm: The totality.

Krishnamurti: Totality. Isn't there a statement in the Jewish world, "Only God can say I am."

Bohm: Well, that's the way the language is built. It is not necessary to state it.

Krishnamurti: No, I understand. You follow what I am trying to get at?

Bohm: Yes, that only this movement is.

Krishnamurti: Can the mind be of this movement? Because that is timeless, therefore deathless.

Bohm: Yes, the movement is without death; insofar as the mind takes part in that, it is the same.

Krishnamurti: You understand what I am saying?

Bohm: Yes. But what dies when the individual dies?

Krishnamurti: That has no meaning, because once I have understood there is no division...

Bohm: ...then it is not important.

Krishnamurti: Death has no meaning.

Bohm: It still has a meaning in some other context.

Krishnamurti: Oh, the ending of the body; it's totally trivial. But you understand? I want to capture the significance of the statement that there is no division; it has broken the spell of my darkness, and I see that there is a movement, and that's all, which means death has very little meaning.

Bohm: Yes.

Krishnamurti: You have abolished totally the fear of death.

Bohm: Yes, I understand that when the mind is partaking in that movement, then the mind is that movement...The mind emerges from the ground as a movement, and falls back to the ground; that is what we are saying.

Krishnamurti: Yes, that's right. Mind emerges from the movement.

Bohm: And it dies back into the movement.

Krishnamurti: That's right. It has its being in the movement.

Krishnamurti: Quite. So what I want to get at is, I am a human being faced with this ending and beginning. [This] abolishes that.

Bohm: Yes, it is not fundamental.

Krishnamurti: It is not fundamental. One of the greatest fears of life, which is death, has been removed.

Bohm: Yes.

The extended dialogues of David Bohm and J. Krishnamurti are clearly precedent-setting, since they paired a Nobel-Prize-winning Western physicist with a well known Eastern mystic and spiritual leader. These dialogues examine deeply a wide variety of human beliefs and expressions, which include the limitations of thought, awareness, existence, truth, death, and the future. I will list them and suggest they be reviewed with contemplative attention which would be consistent with the spirit of Bohm and Krishnamurti (Keepin, 1994.)

- There is a veritable chasm between reality and truth. They are certainly not the same thing. While illusion and falsehood are part of reality, they are not part of truth.
- Truth is one; reality is multiple and conditional.

- Truth transcends and comprehends reality, but reality does not comprehend truth. Reality is every thing, but truth is no thing.
- Our minds are occupied with reality, when we really need only truth. The security we seek does not reside in reality or thinking but only in the nothingness of truth.

These insights were characteristic of Krishnamurti and supported Bohm's conviction that awareness lay beyond thought and was the very source of true creativity, insight, and intelligence. Bohm met with the Dali Lama as well as many other prominent spiritual masters as he pursued questions and answers beyond science, searching for new insights and clues.

The influence of metaphysical concepts is quite visible in Bohm's later work, especially in the developmental of his idea of the superimplicate, which is the deepest and most subtle of his formulations. Bohm was the first to incorporate spiritual values in physics, and to provide metaphysical and philosophical dimensions that interface and optimize stringent scientific and empirical thought (Keepin, 1994). Free-form dialogue was particularly attractive to Bohm as a way of going beyond the usual thought structures and content. He said that thought had the intrinsic disposition to divide things up and inferred that thought was not a powerful contribution to the truth (Bohm, *On Dialogue,* 1992).

In free-form dialogue, one places opinionated positions on hold while considering the alternatives expressed by others. This allows for the generation of new insights in an expanded collective consciousness. This kind of refined dialogue is still being practiced at the Massachusetts Institute of Technology (Isaacs, 1993).

In concluding this chapter, we will consider some of the comments of Ken Wilber, who has had an ongoing involvement with interpreting, supporting, and assessing dialogue between scientists and metaphysicians. While supportive of this interchange, Wilber, in 1982, made the following statement:

> I suggest that the new physics has simply discovered the one-dimensional interpenetration of its own level (nonsentient mass/energy). While this is an important discovery, it cannot be equated with the extraordinary phenomenon of multidimensional interpenetration described by the mystics.

> We saw that Hinduism, as only one example, has an incredibly complex and profound theory of how the ultimate realm generates the causal, which in turn generates the subtle, which creates the mind, out of which comes the fleshy world, and, at the very bottom, the physical plane. Physics has told us all sorts of significant things about that last level. Of its predecessors, it can say nothing (without turning itself into biology, psychology, or religion). To put it crudely, the study of physics is on the first floor, describing the interactions of its elements; the mystics are on the sixth floor, describing the interaction of all six floors.

Many physicists and scientists have taken the philosophical position that the world is hierarchically structured, where each higher level contains the lower but not vice versa. In a conversation with Renée Weber, David Bohm made this most integrating comment: "First of all, the hierarchical position seems to reject the notion of the immanence of the whole. To my mind, ancient tradition includes both immanence and transcendence. Certainly the Buddha, and a great many other philosophical and religious teachers as well would agree with this." (Weber, 1986).

In the early 1980s, Wilber made the following insightful comment, which has proved to be prophetic:

> Agree or disagree with the new paradigm(s), one conclusion unmistakably emerges; it makes ample room for spirit. Either way, modern science is no longer denying spirit. And that is epochal.

As theologian Hans Küng remarked, the standard answer to "Do you believe in spirit?" used to be "Of course not, I'm a scientist," but it might very soon become "Of course I believe in spirit. I'm a scientist."

Summary

Science and mysticism are both valid approaches to nature. However, Lawrence Beynam, in 1978, commented, "We are currently undergoing a paradigm shift in science...perhaps the greatest shift of its kind to date." Similar to quantum physics, metaphysics has come to state that underneath observed reality is another order of reality that is unifying, undivided, peaceful, and eternal. Science has searched for a grand unitary theory for everything, whereas mysticism is the experience of oneness with everything. While science has strived to explain the mystery of matter and being, mysticism has strived to experience it. Science and mysticism have grown closer and closer in this quantum and post-quantum era. Bohm and Krishnamurti became aware that science, like mysticism, was part of the spiritual process—absolutely necessary for expansion of our consciousness.

Questions and Answers

Q: Do you believe that access to the collective unconscious really makes a contribution to the dialogue process?

A: There is considerable scientific support for the existence of a collective unconscious as described by the Swiss psychoanalyst Carl Jung. Second, I see no reason why it should not play a part in an investigational awareness seeking dialogue. Any consciousness-joining process with another is decidedly going to provide an opportunity for the two or many to have a context that can go beyond what each person might accomplish on his own. Since all minds are joined at all levels, dialogue can invite a willing suspension of belief in perception and causality and an openness to unity and a divine presence. Intentionality, respect, and loving regard for the other greatly facilitate insight beyond ourselves and can give rise to the reestablishment of a collective consciousness. I would also regard psychoanalytic treatment (with the merging of the unconscious of the analyst and analysand), in the service of making the unconscious conscious, as a time-proven delimiting example of collective unconscious presence.

Q: Does the unified field theory have an application in the dialogue process?

A: The dialogue in this chapter has as its orientation a search for unity and therefore intentionally embraces the unified field theory of reality. Its intent is to align itself with the strong attractor fields of meaningful reason, reverence, love, joy, peace, and enlightenment. This unconscious and conscious alignment employs a sustaining power rather than unsustainable force. Conventional problem solving moves from the known to the unknown; however, in open dialogue oriented toward illumination, enlightenment, and pure consciousness, just the opposite occurs. In this altered paradigm of dynamic nonlinearity, one moves from the unknown to the known. The unified field theory, joining, and dialogue are lucidly expressed in a passage from *A Course in Miracles*: "Alone we can do nothing, but together our minds fuse into something whose power is far beyond the power of its separate parts. The kingdom cannot be found alone, and you who are the kingdom cannot find your self alone."

13

Science and Mysticism

Legacy of the Mind, Legacy of the Spirit

The softest things in the world overcome the hardest things in the world.
Non-being penetrates that in which there is no space.
Through this, I know the advantage of taking no action.

—Lao-tzu

Now let us consider how metaphysical mysticism can make a most valuable contribution to our progress toward choosing true reality and a higher consciousness. In 1975, Renée Weber posed this very important question for us to consider: Are the good, the true, and the beautiful attributes of the universe? An analysis of mysticism reveals that timelessness, beauty, elegance, aesthetics, simplicity, wisdom, truth, ethics, and good are all essential qualities inherent in mystical/spiritual consciousness. It was her observation that in the twentieth century, ancient Indian and Platonic metaphysics converged with modern field physics. In Plato's Philebus he commented, "then, if we are not able to hunt the good with one idea only, with three we may catch our prey: Beauty, Symmetry and Truth are three..." In the

Brihadaranyaka Upanishad there appears this comment: "The law is the Truth and the Truth is Law. The Truth and the Law are one."

The transcending affirmation that goodness, truth, and beauty are attributes of the universe is a legacy left to us by the sages of Greece and the seers of India. These attributes find their most succinct and lucid expression as a composite of Plato's idealistic cosmology, metaphysics, and ethics.

Unexpectedly, quantum physics has arrived at conclusions precisely parallel to Plato's earlier view of the cosmos, as represented by truth, beauty, and good. Steven Weinberg, a Harvard theoretical physicist, holds an essentially Platonic view of the universe. He points out that the essence of subparticle physics is the aesthetic and is derived from the concept of beauty.

Not surprisingly, he makes the observation that the nonsymmetrical relationships that exist in particle physics are displaced by the greater underlying symmetry and harmony of subparticle quantum mechanics and quantum reality. In the process of explaining our perception of the material and the laws of particle physics, Weinberg commented in 1974, "The answer...is that superficially some of these symmetries are broken and that nature, as we observe it, is but an imperfect representation of its own underlying laws."

Physicist Henry Margenau stated in 1961 that elegance and simplicity are aesthetic requirements of a valid physical theory. Earlier, in 1938, Albert Einstein said, "What is impenetrable to us really exists, manifesting itself as the highest wisdom and the most radiant beauty." In 1967, Richard Feynman, Nobel Prize laureate in physics, addressed the issue of choosing the correct theory from a number of diverse scientific theories, by explaining that we can choose the one theory that is "right" and "true" by recognizing its simplicity and beauty. Feynman made the following reflective comments: "To those who do not know mathematics, it is difficult to get across a real feeling as to its beauty, the deepest beauty of nature." In 1964, addressing the problem of extrapolation (to infer a value from within an already observed interval) and prediction, a distinctive characteristic of science, Feynman wrote, "What is it about nature that lets this happen, that it is possible to guess from one part what the rest is going to do? ....I think it is because nature has a simplicity, and therefore a great beauty." He perceived that beauty has consequently acquired predictive ability. When compelling theories are evaluated, it serves to guide science toward the truth.

We are now going to place the ideas of beauty, truth, and good in a historical context, so we may more fully appreciate the fact that they have already been our intellectual and philosophical heritage since time began.

In the fourth century BC, Plato formulated his concepts of beauty, truth, and good using the dialectic method of inquiry; here discussion and dialogue became the way of intellectual investigation (*Philebus*, *Phaedo*, *Timaeus*, *Theaetetus*, and the *Republic*). This method provided Plato with an unmediated, intuitive insight. David Bohm also turned to dialogue as a principal means of intellectual inquiry, just as Plato had done in the fourth century BC.

Plato deemphasized the world of perception, citing its unrelated randomness as opposite to his concept of reality. The reality of Plato was eternal, ultimate, and immutable. Furthermore, it was nonmaterial, unalloyed, simple and therefore without dimension, incapable of being pulled apart by time. He maintained that knowledge would be impossible without truth. Without truth, the observer's opinion and beliefs would be incapable of dispute, and only separate realities would prevail. Without truth, no claim could be proved false, and consequently, non-true, leaving only opinion. Plato held this position to be untenable (*Theaetetus*).

Truth as formulated by Plato involved cosmic principles, whereas in the perceptual world there is only a muted likeness of its source. In the non-perceptual world, the truth and being of everything is indissolubly connected to the truth and being of the whole. This concept provided for a basic unity of being.

Plato conceptualized beauty as not sensuous, but rather as abstract, requiring no social consensus, being a cosmic concept connected to truth and goodness. He stated, "The power of the good has retired into the region of the beautiful, for measure and symmetry are beauty and virtue the world over." (*Philebus*).

Foreshadowing twentieth-century mathematical physics, Plato proclaimed that "truth is akin to proportion" (*Republic*). He thought mathematics was an insufficient inquiry into the region of being to which dialectic alone could lead us. He commented that "numbers are...a thing...which I would...call useful: that is, if sought after with a view to the beautiful and good; but if pursued for any other reason, useless." (*Republic*).

Platonic metaphysics cannot be grasped without the idea of good. Although Plato repeatedly affirms that goodness is inseparable from beauty and truth, he does not provide criteria for good as he does for beauty and truth. Plato does tell us how good functions and what its absence portends.

Good is clearly no mere human sentiment or expression of approbation or taste, but rather a force of nature. In the *Republic* he characterizes good as "that which imparts truth to the known and the power of knowing to the knower." As to the objects of knowledge he states, "these derive from the good...their very being and reality." Nevertheless, the good cannot be equated with being, for it exceeds it, for good is the "author of science and truth, and yet surpasses them in beauty."

Like the sun in our visible universe, the good gives being and life to all entities, for if we do not invoke a force of cosmic cohesion and meaning, we are left with pseudo explanations instead of a genuine theory of nature.

Plato felt that value was therefore woven into the very fabric of creation, since good furnished divine ideation with its model. The perfection of the good was translated into the imperfect and refractory medium known as matter, but only representing a mere reflection of its source.

A reading of Plato provides a mounting certainty that good is not merely a metaphorical or mythical device, but something we can directly experience. "In the world of knowledge, the last thing to be seen and only with great difficulty, is the essential form of good...without having had a vision of the form, no one can act with wisdom...the soul...can learn only by degrees to endure the sight of being, or in other words, the good" (*Republic*).

While Plato imparts a conviction that being is saturated with good and the cosmos is perfect, man regrettably has a great tendency to see beauty, truth, and good only on the level of his perceptual world.

Whereas Plato and Aristotle used dialectic (discussion and reasoning by dialogue) as a method of intellectual investigation, in ancient Indian philosophy the acquisition of knowledge led away from the consciousness of sensory experience to a pure consciousness, where consciousness became the equivalent of the object it seeks. Indian philosophy employs meditation to reach a state without duality and devoid of "self-nature." Both Indian and Tibetan philosophy view meditation as a form of empiricism (the theory that all knowledge originates in experience)—free from unwanted assumptions, radical in the sense of reverting to the roots of experience. The primary aim of both dialectic and meditation is to experience the principle of being and source of life. This immutable principle of reality embraces the concept of unity, expressed as the "ultimate one."

The Platonic concepts of truth, beauty, and good find their correlation in Eastern philosophy: *cit* represents truth or consciousness; *ananda,* beauty or bliss; and *sat,* good or being. We also find attributes of Platonic reality in ancient Indian philosophy. The Indian concept termed *Brahman* represents the ultimate principle of reality, expressing the concept of the one rather than many; abstract and universal rather than concrete and particular; beyond words, thoughts, and concepts; ideal and therefore nonmaterial; eternal, therefore uncompounded and unborn; infinite and nonlocalized, therefore *field* in nature.

Both ancient Indian and Platonic philosophy address the source of being and the ultimate principle of life; the reality of these two philosophies interfaces with the unified field theory of quantum physics. All three state that underlying reality consists of unity, harmony, nonmaterial intelligence, and timelessness. All three entities argue that our perceived multiplicity (fragmentation and separateness) is an illusory figure superimposed by the *unenlightened* consciousness upon the one true, undifferentiated ground and that this true ground is space itself; its pure "nothingness" seethes with activity and creates something out of itself alone. Physicist John Wheeler—in discussing beauty, truth, and good as qualities of superspace—stated in 1975 that anyone who accepts the quantum principle is forced to believe them. Indian and Platonic philosophy, and quantum physics, state that a change in consciousness is necessary to appreciate and experience the highest realities, the spiritual and eternal essence of the real: again, beauty, truth, and good. (Weber, 1975).

Quantum physics has begun to join Platonic and ancient Indian philosophy in recognizing that the consciousness that asks for an answer to the question of creation exists in time and cannot therefore understand the answer. For the enlightened consciousness, which alone could comprehend the answer, the answer does not exist in time, and as a result, the question disappears.

In 1986, Weber posed several questions that are crucial to this inquiry of whether truth, beauty, and good are properties of nature and whether they can be validated by science. They are:

- Is it legitimate to ascribe properties like truth, beauty, and goodness to nature as a whole?

- Is there some additional feature of truth (absent in the scientific vision, but present in that of sage and seer) that would suggest why science has not yet succeeded in replacing these ancient claims?
- Since science accepts the concepts of truth and beauty, can we, extrapolating from philosophical wisdom, predict that it will ultimately confirm the good?

After extensive evaluation, Professor Renée Weber considered truth, beauty, and good to be legitimate properties of nature, citing the supporting wisdom and knowledge of both Platonic and ancient Indian metaphysics.

Science has not held itself competent to determine good, placing it in the category of "value." Once science took on the position that only measurement is objectivity, it disqualified itself from any official pronouncement on the concept of good. This decision has had the unfortunate result of not only discouraging inquiry into the existence of good, but also placing it in the category of scientific illegitimacy. This boundary restriction of science has resulted in a denial of the ontological[3] reality of good, placing science in the arena of neutral, cold, and dehumanized ideology.

If meaning becomes separated from beauty, truth, and goodness—its supreme principles—then meaning becomes weightless and unable to claim man's allegiance, inspiration, or even his interest. The result is a sense of unconnectedness to the truths science exhibits. Since science already accepts the concepts of truth and beauty, with good being their inherent corollary, Weber foresees that science will eventually accept good as the companion to truth and beauty.

In the meantime, Weber notes that the metaphysicians and sages have for thousands of years comfortably plied the waters of truth and beauty as well as good. She suggests that science might use radical empiricism, which could lead the scientist to use himself as the test object (in much the same way the mystic uses the self as the test object) to determine whether or not this third attribute, "good," exists. She conjectures that if this experienceable radiance, "good," which according to Plato is so powerful that it "makes truth possi-

3. Based upon being or existence.

ble," ever becomes part of the domain of scientific discovery, we can anticipate a revolution in thought that will redraw the map of being. For if truth, beauty, and goodness are one, then fact and value become one, and inner and outer become one. With this transformation in consciousness, the false, fragmented, schizoid world we live in would be healed.

The Indian poet Rabindranath Tagore, a Nobel Prize laureate in literature, wrote a poem in 1920 that eloquently expresses the transition from the explicate order to the implicate order:

> The same stream of life that runs through my veins, night and day, runs through the world and dances in rhythmic measure. It is the same life that shoots in joy through the dust of the earth into numberless blades of grass, and breaks into tumultuous waves of leaves and flowers. It is the same life that is rocked in the ocean cradle of birth and death, in ebb and flow. My limbs are made glorious by the touch of this world of life and my pride is from the life throb of ages dancing in my blood, this moment.

It appears that the most important message of existential and cognitive psychology is that we do have the ability to make the choice to correct our erroneous mental cognitions. The most important message of quantum mechanics is that it provides the corrected cognitions that represent unity, wholeness, and infinite harmony as the real reality of consciousness based upon being or existence.

In our current state of consciousness, we have to choose these correct cognitions again and again until they completely supplant the distorted and inaccurate ones. In becoming conscious and responsible, we must discover and develop the meaning of our existence. Absence of meaning removes any inclination to see that a choice is available. Considering all of the foregoing, the absence of meaning can only be removed by finding the unity and wholeness that are beyond time and space, for human strength is importantly sustained or defeated depending on the awareness of an inner core of presence or spiritual strength.

Ancient and contemporary mysticism have provided us with the following direct insights into the process of moving from perceived reality to true reality. Mysticism points out that the mind has shown us a dream, a dream of observed separation and fragmentation from the knowledge of spirit. Further, the mind has equated itself with the dream. However, we are not our dream, but are the dreamer. As we look knowingly about, we will see everyone having the same projected dream of fear, attack, separation, and guilt. When we fully recognize this, we can identify with our companion dreamers and take the first necessary joining step that will move us toward a different reality. We are all separated by differences in form, but we are all joined by our shared content...unity. The dream of our separation introduced degrees, intervals, aspects, and components. Since a dream is not real, these divisions make no sense and can only result in internal conflict with its senseless comparisons, judgments, fears, and guilt.

The dream that has created separation has also created fear. This fear is projected out, creating a world of illusion—an illusion of material and its inevitable destruction. Again, projection, that very early mechanism of consciousness, reflects only what is within. The dream has caused energy to be reflected backwards (introjection), maintaining a reduced level of consciousness.

Projection once again has created perception. Prior to the dream of separation, perception did not exist. Mysticism believes that we are creations and extensions of "the Supreme Spirit of the Universe." This Spirit did not cooperate and agree in the choice of our dream, but remains in unity and love.

Before ending this chapter, we will return to Ken Wilber's ladder of consciousness:

1. Ultimate: consciousness as such, the source and nature of all other levels
2. Causal: formless radiance, perfect transcendence
3. Subtle: archetypal, trans-individual, intuitive
4. Mental: ego, logic, thinking
5. Biological: living (sentient) matter/energy
6. Physical: nonliving matter/energy

The lower four rungs of the ladder—the physical, biological, mental, and subtle—are expressions of material energy. The upper two rungs of the ladder—the causal and the ultimate—are expressions of nonmaterial intelligence, perfect consciousness, and unity. The gradient of consciousness clearly flows toward its highest level. Upon further examination, one finds another associated gradient where effect flows toward cause. Expressed another way, ideas and thoughts are joined ever more closely with their source. On the lower four rungs of Wilber's ladder, material and energy are expressions that are projected onto the world's "screen," appearing as images of division and separation. Here, material becomes the expression of ideas that have left their source. These fragments—representing form, change, and death—are effects without cause. Therefore, they are illusions and are not real. On the uppermost rung of the ladder—the ultimate—cause and effect have eternally and joyously become one.

The dream of fear, guilt, and lack has become projected onto our creation of a distorted false supreme spirit that like us is separated, fearful, vengeful, and attacking. It has been said that only the self-accused can condemn. Once again, this process traverses the cycle of projection and perception. Voltaire commented, "God has created us in his image and likeness and man has returned the favor." This protective, introjective process has provided man with a sense of immense guilt and foreboding. In this process, projected illusion has become our hiding place from our projected image of a wrathful deity. This projected vengeful, jealous deity is also an illusion and is therefore causeless.

Summary

Mysticism points out that since man is the dreamer and not the dream, we can choose again. We can join our brothers and choose a happy dream of peace and love, which can lead us back to where we have never left! We have forgotten that Spirit created us and that like Spirit we are creators, not projectors or perceivers. We were created by Spirit, and obviously as Spirit we need suffer nothing.

Questions and Answers

Q: The illusory self, or ego, is so pervasively shared and feels so real, it is very challenging to view it otherwise. Could you reconceptualize it so that viewing it as truly false could be seen more lucidly?

A: The ego has maintained the illusion that we have created ourselves by way of our bodies and then projected the mental images out onto the world screen, where they can be perceived. This process leaves us in a position where we regard ourselves as separated from the process we effected in the first place, and we now believe we are acted upon by a world beyond our control. Reframing this, we believe we are split off from ourselves and split off from our divine source. The ability of the ego to perpetuate itself by diverting us from mindfulness to mindlessness should never be underestimated. We did not create ourselves; we are apt to forget this. David Hawkins, in his book *I: Reality and Subjectivity* (2003), states, “the ego is the victim of itself. With rigorous inspection, it will be discovered that the ego is really just ‘running a racket’ for its own fun and games and survival. The real ‘you’ is actually the loser. The ego has an endless store house of prizes on which to feed. It greedily pounces on sentimentality, on the virtue of being right, on the prize of being the victim, or on the martyrdom of loss and sadness. It also offers the excitement of winning or gain as well as the pain of frustration. It offers the ego inflation of getting attention or sympathy. One can see that each emotion is, in and of itself, its own pay off. The ego clings to emotionality, which is intimately connected with its positionalities; it pretends to think it has no choices. To ‘surrender to God’ means to stop looking to the ego for solace and thrills and to discover the endless, serene joy of peace. To look within is to find the underlying, ever present Source of the illumination of the mind itself.” There is a choice: One can forsake the ego’s forced forgetfulness, denial, arrogance, and deception and intend to leave the world not by death but by *truth*.

14

Psychology, Physics, and Now, Mysticism

Bridges to Beauty, Truth, and Good

What is needed today is a new surge that is similar to the energy generated during the Renaissance but even deeper and more extensive…the essential need is for a "loosening" of rigidly held intellectual content in the tacit infrastructure of consciousness, along with a "melting" of the "hardness of the heart" on the side of feeling. The "melting" on the emotional side could perhaps be called the beginning of genuine love, while the "loosening" of thought is the beginning of awakening of creative intelligence. The two necessarily go together.

—Bohm & Peat

Clear as Your likeness does the light shine forth from everything that lives and moves in you. For we have reached where all of us are one and we are home where You would have us be.

—*A Course in Miracles*

Throughout this book, we have examined man's use and understanding of psychology, physics, and mysticism. Now we will examine the interconnections between these three entities and their inherent relationship to the emergence of highest consciousness, our true reality. Beginning with psychology, we will comment on projection. Projection, that most important mechanism of mental functioning, can be considered an early emerging property of consciousness. The inclusion of projection as a function of consciousness is clearly supported by the currently amended and expanded view of consciousness as unitary. Viewed in this light, it should be no surprise that the mechanism of projection can be seen, suffusing as well as interconnecting psychology, physics, and mysticism. An understanding and appreciation of projection and its apparent movement signals to the observer that genuine movement and expansion of consciousness lie hidden behind it. Without close observation and examination, projection and its counterpart introjection can portray reality, but it will be false, for they can only mimic true extension and expansion of consciousness. The fact that projection is false and generates illusion rather than true reality is clearly supported by the fact that projection possesses the goal of division, separation, and fragmentation rather than the goal of joining and unity.

As was mentioned earlier, existential psychologists make a serious case for the abolition of the idea of an unconscious, folding it into what they feel is a more functional and supervening concept, seamless consciousness. The existentialists view the unconscious as an equivalent form of self-deception based on volition.

Cognitive psychology, while not directly refuting the concept of the unconscious, functions as if unconscious constraints are not in place. It accesses feelings and cognitions that have appeared unavailable for recall, utilizing motivation and focused attention. Much of what this system retrieves can be considered as coming from the realm originally designated as the unconscious. The concepts and methods of cognitive psychology have been extraordinarily successful in correcting unrecognized, dysfunctional thinking and behavior; in fact, the success of this method is unparalleled in the history of psychology and medicine.

Several bodies of psychological thought have taken the position that the unconscious does not exist, for consciousness is seamless. I strongly believe there is another more helpful way to look at the conscious/unconscious paradigm. First, I would agree that consciousness in the broad sense is seamless. However, enfolded in it are three, if not four, entities: the conscious, the preconscious, the unconscious, and the fourth, which I will call the super-unconscious. In the conscious mind, awareness can access all that is stored in memory. In the preconscious, memories that are not available for recall can be accessed by mental effort. In the unconscious, certain affects (feelings), primitive drives (sexual and aggressive), and memories are not available for recall even with effort, because they are too fearful or repugnant to the conscious mind. They are held out of awareness by the automatic mental mechanisms of denial and repression. They can be accessed only by free association, dream analysis, or in rare moments of lucid thinking, but only if their fearfulness has been sufficiently neutralized. In psychoanalysis, neutralization of fear is a central part of the treatment process.

The dysfunctional effects of this hidden material can also be corrected by using a cognitive therapeutic technique that challenges one's belief in the need to hold on to the effects by demonstrating that they do not have a true cause.

Material that is hidden in the super-unconscious mind is also kept out of awareness by a process of denial and repression; however, super-unconscious processes are far more organized and subtle than those that operate in the unconscious mind. To maintain its power, the super-unconscious extensively employs quasi-quantum principles of nonlinearity and nonlocality. These quasi-quantum principles are false, not true, because their intent is to promote separation and division. The true intent of quantum mechanics is always *unity*. The reason that super-unconscious material is so powerfully defended against, is that it is far more fearful than anything at the unconscious level. Hidden there are ideas or thoughts that we have created about ourselves, our world, and our universe, that have displaced God. Here, effect is regarded as cause. In other words, here we have effect disassociated from cause, which is in fact illogical, for the effects are causeless. For example, at this level we believe that while we created ourselves and our universe and now we are acted upon by things outside ourselves (we become victims of circumstances) so here we regard cause as effect and effect as cause. In a word, this is decidedly psychotic.

This whole system is kept in place (out of awareness) by immense fear. The fear is that if we become aware of this repressed material, God will destroy us. This is another illogical assumption, because as a creation of God, we can only have the attributes of our cause—*love*. The logic of true cause and effect dictates that thoughts cannot leave their source. The goals of both the unconscious and super-unconscious are diversion, separation, attack, and destruction, and the maintenance of these and their associated delusional ideas.

Highest quantum and mystical consciousness states repeatedly and clearly that pain is illusion, sleep, and deception; and joy is reality, awakening, and alone the truth.

This idea leads us back to a reconsideration of cognition. Cognition can be described as a state of knowing, derived from the systems of information that we have accessed and activated. Providing we are integrating data clearly, and projecting from a point of love and empathy, consciousness expands. These information systems generate consciousness as we ordinarily observe it. Cognition and projection occupy primarily that arena of observed order termed explicate. This type of order is primarily unhidden and accessible, and it incorporates local causality. Defined further, it is decidedly locally constructed and does not have quantum mechanical capabilities.

This explicate consciousness projects whatever the cognitions are—correct or incorrect—to the outside of the mind, where they are experienced as perceptions of division, separateness, fragmentation, and chaos. In this kind of perception, the second law of thermodynamics prevails, that of entropy. The law of entropy states that the entire universe is trapped in a process of dysfunction, fragmentation, disintegration, illness, and finally death.

In contrast to the unfolded explicate order of consciousness, the enfolded implicate order of consciousness is a reality of timelessness, harmony, wholeness, nonmaterial energy, and intelligence. In essence, this reality represents just the opposite of the law of entropy. This underlying order represents the true and unchanging reality of unity, while the explicate order of consciousness represents divisibility and separateness in time and space, and connection only by local causality. Here, illusion prevails.

Most humans, while they are alive on the planet, live in this explicate order of consciousness. Put another way, most humans live in a vast perceptual illusion generated within their own minds and projected outward. If our collective thoughts have created this physical world, then it would follow logically that changing our collective thoughts can change the world!

One question that you might ask would be, *Why* is it mankind has remained focused on this explicate illusion of reality, when there is an underlying reality of timelessness, peace, and harmony? The cognitive errors that have supported this distorted, disastrous perception of the world have their basis in mankind's stable illusions of chaos and guilt, and his conviction that there is never enough of what he thinks he needs.

At this point, we are going to summarize and compare the traditional physical views (old conscious cognitions) of the world with modern physical views, as presented by relativity and quantum physics (new conscious cognitions), (Chopra, 1985). It can be noted that the old concepts conform to Bohm's explicate order; while the new concepts conform to his implicate order.

As we follow the transformations from traditional views to modern views, we can become aware of the corrective cognitions and extensions (projections) that are central in our ever-expanding consciousness. More specifically, in examining the following cognitions you will note that the traditional, or old, cognitions are characterized by projections that fold back and return perception and consciousness to the mind of the sender (introjection) without any addition or expansion. The new, or quantum, ideas and cognitions are characterized by an ever-increasing expansion of consciousness. As these cognitions move toward highest consciousness, projective/introjective mechanisms gradually fade out and are replaced by unlimited creative expansion. We have thought of the world out there as independent of us, when in fact it is a construction of our own interpretation.

The twentieth century presents us with many important and helpful paradigm shifts in consciousness. These paradigm shifts were generated by quantum physics experiments, theory, and metaphysical thought.

For classification and understanding I will list old paradigms and their recent shifts. Time has been regarded as an absolute with a linear progression. The new paradigm states that time is not absolute but an illusion resting on the eternal present. Time passing cannot be scientifically measured, since there is only an eternal present. (Wapnick, 1990). The older paradigm recognized the material world and people as particles of matter separated from each other by time and space. The new paradigm pointed out that while we might have the sensations of matter they are in fact waves of energy and form in a void of energy and form. Empty space is not a designated emptiness of nothing but a repository of non-material intelligence. The new paradigm states that the world is non-material and is composed of energy fields that are derived from one manifest underlying field. Space and time are also an unfolding of this one field.

The old paradigm said that thinking does not make it so. The new paradigm states that all true thought first begins as mental image and then extends as unlimited meaning and energy. The new paradigm states that at a very deep level of universal consciousness there is formless information that creates form by organizing energy in a pattern where it expresses properties of matter.

The old paradigm held that our objective world (which is out there) is entirely independent of an observer. The new paradigm indicates that this is a participatory universe where the objective world is a direct response to our observation.

In the old paradigm mind and matter are separate and independent, whereas the new paradigm states that mind and matter are identical. The subjective experience of the unified field is our mind where the experience outside ourselves is one of material objects.

The old paradigm told us that the mind was imprisoned in the brain and our intelligence in the nervous system. The new paradigm presents information that mind and intelligence are not imprisoned by the body but extend out as energy fields in a limitless manner to the universe and beyond. The perception of localized forms is only one expression of the mind and does not negate its limitless infinite essence.

The new paradigm identifies the entire cosmos as a field of intelligence where thinking and interpretation provide us with material perceptions. Human beings are focal points in this extending intelligence field where the focal points represent organized relationships.

Finally, the old paradigm represented consciousness as an expression of matter. The new paradigm regards matter as an expression of consciousness. With this concept in view one could say that bodies are lower consciousness, robotic expressions that have been imbued with thinking. Matter and thinking physical bodies are in fact very distant expressions of the underlying manifest field of highest consciousness. This concept is almost memorialized in David Bohm's definitive comment that thought is not an effective pathway to truth.

Our old paradigm presented energy as both limited and fragmented. Our new paradigm says that energy is unlimited, unbounded, and flowing, characterized by a new dimensionality of wholeness, nonjudgement and compassion. Physics and ethics have now become one in this process, for the limitless energy of this whole is somehow bound up with what we call spirituality (Chopra, 1985). Today, many quantum physicists state that this energy itself is love. Mystics have said, "how can it be anything else?"

"So you're little Bobbie; well, Rex here has been going on and on about you for the last 50 years."

Realization of final reality

When you list all of the new paradigms together, as we have, they radiate a radical quality that can be overpowering, in spite of their spectacularly positive and reassuring messages. This radical quality generates an immense fear of acceptance that can only be diminished by a continuing attitude of quiet, thoughtful, and peaceful consideration.

The emerging new paradigms of the twentieth century are particularly important for they reflect an early organized paradigm and hologram of higher consciousness. In the first part of the Twenty First Century the planet continues to manifest pervasive dysfunction and chaos. However, these breakthrough paradigms indicate that the planet is beginning to slowly move into the attractor fields of unity, truth and good. These attractor fields send out the powerful message that the laws of matter do not apply to the reigning and underlying field of greatest intelligence and consciousness.

In the old paradigm of consciousness we can have only limited consciousness which is projected out and then taken back unaltered. Within the new paradigm, consciousness is ever expanding gradually displacing projective/introjective mechanisms.

These laws of matter are connected to time, space, and causation, which are not real, but rather are illusions of the explicate world of sensation and perception. One's real essence is this field of awareness, which interacts with its own self and becomes both mind and body. Again, the new quantum physical view states that humans are basically consciousness, which then conceives, governs, and actually becomes the body and the mind; this in turn generates and orders our perception of the world and the universe. Again, this process occurs by way of projective mechanisms that can cause consciousness to flow either forward or backward, depending on how enlightened our cognitions are at that point. As consciousness flows forward, projection phases out and is eventually replaced by knowing.

The old paradigm said that energy was both limited and fragmented, while the new paradigm says that energy is unlimited, unbounded, and flowing, characterized by a new dimensionality, wholeness, nonjudgment, and compassion. Physics and ethics have now become one in this process, for the limitless energy of this whole is somehow bound up with what we call spirituality. Today, many quantum physicists state that this energy itself is love. Mystics have said, "How can it be anything else?"

When you list all of the new paradigms together, as we have, they radiate a radical quality that can be overpowering, in spite of their

Old Paradigm	New Paradigm
Linearity and locality prevail.	Nonlinearity and nonlocality prevail.
Time is an absolute.	Time is only a thought or an idea. There is only one eternal presence.
Material and bodies are made up of particles of matter separated by time and space.	The whole material world is made up of particles that give a sensation of matter through our senses but are really fluctuations of energy and form in a void of energy and form.
Matter is created from form.	At a deep level, there is a dimension of information that creates form out of formlessness by influencing the behavior of matter, placing it in a formation that will exhibit properties of matter.
Empty space is nothingness.	Empty space is not nothingness, but rather a fullness and reservoir of nonmaterial intelligence.
Human beings are self-contained, separate entities.	Human beings are not self-contained, but rather are focal points in a field of intelligence. Our real essence is this field of intelligence that interacts with itself.
Mind and matter are independent and separate entities.	Mind and matter are essentially the same and are participatory. The objective world is a response to the observation we've created.
We can have only limited consciousness, which is projected out and then taken back unaltered.	Consciousness is ever expanding, gradually displacing projective/introjective mechanisms.

spectacularly positive and reassuring messages. This radical quality generates an immense fear of acceptance that can only be diminished by a continuing attitude of quiet, thoughtful consideration.

Upon considering the pervasiveness and the tenacity of the erroneous and grim ideas that our minds hold to be real, we can conclude that it is small wonder that they have obscured and shrouded any positive, deeper levels of reality. In order to truly appreciate the power of their mechanisms, let us revisit just a few of these widely held false beliefs:

- Happiness depends upon the acquisition of material things.
- My happiness depends upon what others think of me.
- There is always a correct and perfect solution to our problems and if it is not found, catastrophe will follow.
- Not getting what I deserve means that I am a failure.
- Each negative event is an indication of a never-ending pattern of defeat.
- The material world, including human beings, is made up of bits and clumps of matter, which are separated from each other in space and time and are destined to eventually disintegrate and become nonexistent.
- Happiness is due to circumstances that are external and beyond my control.
- I am not worthy of peace and love.

Cognitive systems and existential psychology have identified our irrational cognitions and the associated irrational fear that if we give them up we will have nothing, be nothing, and literally cease to exist. This aggregate of faulty cognitions constitute what might be called the old paradigm of observed reality and collective consciousness. Quantum physics and metaphysics have now come along to present us with a new paradigm of corrected cognitions that is destined to become our true collective consciousness. The choice is not between something and nothing, but rather between something we have and something we can have, which is far, far better. In fact, it is the choice between the imperfect reality of perception and the perfect realty of true knowing.

Since these old, erroneous paradigms are universally held in consciousness, they have generated mental images that have been automatically projected into and onto the outer world. If we can slowly recognize this process, it will not come as such a great shock that we are the authors of what we see and perceive.

The explicate order of perceived consciousness can be considered an expression of our current, collective consciousness, whereas the implicate and superimplicate orders of reality are guided by the concepts of quantum theory and mechanics, and are associated with true consciousness. If this is all true, which it appears to be, the observed planet and universe as we see them have been generated by our erroneous cognitions—cognitions that are collective and that have gone both backward and forward in time. Einstein, in his quest for the underlying laws of order and unity, proclaimed that "God does not play dice with the universe." Based on the information presented so far, God is not playing dice with the universe, but *man* is.

You may have an opportunity to briefly experience an example of a paradigm shift in expanded consciousness by practicing the following as a meditation. This is Lesson 45 from *A Course in Miracles* (1975): "God is the mind with which I think...My real thoughts are in my mind. I would like to find them." This meditation can be used for three, five-minute practice periods for two or three days, or more. This practice may provide you with thoughts that you would not ordinarily think. I will quote *A Course in Miracles* directly because I think this would be most helpful:

> Today's idea holds the key to what your real thoughts are. They are nothing that you think you think, just as nothing that you think you see is related to vision in any way. There is no relationship between what is real and what you think is real. Nothing that you think are your real thoughts, resemble your real thoughts in any respect. Nothing that you think you see bears any resemblance to what vision will show you.
>
> You think with the Mind of God. Therefore, you share your thoughts with Him, as He shares His with you. They are the same thoughts because they are thought by the same Mind. To share is to make alike, or to make one. Nor do the thoughts you think with the Mind of God leave your mind because thoughts do not leave their source. Therefore, your thoughts are in the Mind of God, as you are. They are in your

mind as well, where He is. As you are part of His Mind, so are your thoughts part of His Mind.

Where, then, are your real thoughts? Today we will attempt to reach them. We will have to look for them in your mind, because that is where they are. They must still be there, because they cannot have left their source. What is thought by the Mind of God is eternal, being part of creation…

Then try to go past all the unreal thoughts that cover the truth in your mind, and reach to the external.

Under all the senseless thought and mad ideas with which you have cluttered up your mind are the thoughts that you thought with God in the beginning. They are there in your mind now, completely unchanged. They will always be in your mind, exactly as they always were. Everything you have thought since then will change, but the foundation on which it rests is wholly changeless.

It is this Foundation toward which the exercises for today are directed. Here is your mind joined with the Mind of God. Here are your thoughts one with His…

One probable cause for the existence of the perceived world is that it is a reflection of our current state of consciousness, an illusory consciousness that strives to be a primary, unique creator apart and separate from true consciousness. This false consciousness, being an illusion, can only create something that is also an illusion: the material world. This is something that true consciousness could not and would not create, for it is like its source…false, valueless, and untrue. The principal thing that explicate consciousness can generate is a universe of material and its strongest wish is to become a first creator, even though this proves to be the ultimate "bad choice." This choice serves to hide, to blot out, true deeper, implicate consciousness. Implicate consciousness cannot create a universe of material because material is an illusion and does not exist at the level of true reality and implicate consciousness; material and material's associated quality, entropy, simply do not exist. At this point, one might ask, Why would explicate consciousness as we usually know it possess cognitions and projections that are associated with entropy and eventual destruction?

If the cognitions and consciousness of unitary and nonentropic wholeness are fully grasped and embraced, then psychologically speaking, our observed perceptions with their explicate order of reality would be threatened with dissolution and extinction! Acting narcissistically, defensively, and fearfully, we have created false, unnecessary cognitions and illusions. These cognitions and illusions have actively distracted and blocked our vision of the enfolded, implicate order of reality. I emphasize the word *fear* because fear and guilt appear to be the primary repressed and denied factors that block the awareness that we are all enjoined at the level of implicate order, with its attributes of ongoing beauty, truth, and good!

Considering the fact that most inhabitants of the planet are firmly entrenched in observed material reality, you can begin to get an idea of how powerfully you are positioned to continue your use of the explicate order of reality and stay connected to its state of collective, fearful consciousness. One only needs to briefly study the history of quantum physics and quantum mechanics to see how truly intensive and extensive our resistance and denial of quantum reality has become.

While human consciousness, at this time, is primarily in the observed explicate order, there is some good news. From time to time, there are certain masses of quantum information that spiral up into the level of our current consciousness, which discloses that an implicate order of reality does indeed exist. These spirals of information often take the form of surprising coincidences (synchronicities). Who of us has not been startled by having the right book or information unexpectedly thrust into our hands, exclaiming, "That's just what I was looking for." In the past, humanity did not pay much attention to these events and information, because of our intense preoccupation with erroneous mental cognitions. These cognitions support our restrictive view that reality is composed of material division, fragmentation, and lack. Considering the incredible magnitude of man's negative misthought, is it really strange that projection would provide a world in which everything is backwards and upside down?

I would like to depart a bit from our current theme, to make a commentary that can be helpful in removing our self-imposed plight: "crap happens." All of our mistaken beliefs conspire to render man's view of himself and his world nearly humorless. As an aside, humor and laughter can be healing and enlightening because their message is, "Maybe things are not as bad as they seem, and maybe there is a more positive way of looking at things." Humor acts to relieve fear, and when fear is removed there occurs a shift to a serene state of consciousness that permits us a glimpse of our underlying deeper reality. Laughter releases the tension around even the heaviest of matters.

At one point during the Cuban missile crisis, Soviet and American negotiators became deadlocked. There they sat in silence, until someone suggested that each person tell a humorous story. One of the Russians told a riddle: "What is the difference between capitalism and communism?" The answer? In capitalism, man exploits man, in communism, it's the other way around." The tactic worked: With the mood more relaxed, the talks were continued. Psychoanalyst Martin Grotjahn, author of *Beyond Laughter* (1958), noted that humor allows for acceptance of weakness and frustration, and while humor may not actually solve our problems, we may discover a way out while we are laughing.

On a deeper philosophical level, it has been said that the spirit of man forgot to laugh at a fleeting and ridiculous wish to displace God and dethrone him as first creator. God had lovingly created man as a full and complete extension of Himself. Man's spirit, attempting to create differently from God, could not do so and could only project a dream world of material, division, destruction, and death. Man's spirit dreamt that he was now "God"; having stolen God's creative power, he projected onto God his own attributes of attack, destruction, and death. He then feared God's revenge, punishment, and eventual annihilation. Man's guilt and fear of God are strongly kept out of awareness by powerful mechanisms of repression that block the awareness that God is only love. Since man's spirit could never actually leave God, his dream and illogical thought system are destined to dissolve in awakening laughter.

Historically, psychology, physics, and mysticism all have had developmental lines of increasing consciousness. The initial orientation of psychology was response dependent, and its focus was symptom removal. Later, psychology came to include the psychobiological and the sociocultural. More recently, individual psychology became more focused on the development of permanent, stable character traits such as consistency, compassion, trust, and empathy. Psychology has become more firmly lodged in a holistic matrix that acknowledges a collective consciousness that is apparent as well as hidden. Man is increasingly seen as living in a participatory way with his material universe, as well as with his metaphysical, spiritual universe.

Physics has traversed a developmental course similar to that of psychology. Initially, the consciousness of physics was confined to the observed material world—a world that could be directly seen, felt, heard, and measured. Further developments saw the emergence of a body of mathematical and creative thought that postulated an underlying reality of unending harmony, unity, and wholeness. Finally, the experiments of Clauser (1972), Aspect et al. (1982), and Zeilinger (1999) provided a verifiable demonstration that nonlocal causality does exist, supporting the existence of a vastly different underlying reality. The unbelievable had moved to the province of the believable. Physics stated that all particles are in continuous contact with all other particles in an information- and consciousness-sharing, nonmaterial, intelligent foam. With the development of quantum physics and quantum mechanics, we have become aware of the mysterious spiritual-metaphysical presence we all share. This presence moves backward and forward in time on its way to an eternal *now.*

The development of mysticism is no exception to the slowly unfolding course of consciousness. Mysticism has always been in the business of the metaphysical, but this intuitive, contemplative process—with its communication and participation in Godliness—originally had a narrower profile. The earlier roots of mysticism were represented by a more isolated, religious, and anthropomorphic identity. In essence, the original forms of mysticism were more reflective of man's human traits, while modern mysticism has undergone a shift toward a more formless presence embracing a common theme of transcendence and enfolded beauty, truth, and good. *That is the essence of true reality.*

A closer examination of psychology, physics, and mysticism reveals that they are all a part of a spiritual process. This conclusion can be drawn because they all involve the mind, which is not in the body, but rather is nonlocal, quantum, and therefore metaphysical/spiritual. We are now at a point in history where these three entities have intersecting beams of awareness that will increase consciousness in a nonlinear way. This process will illuminate a landscape of higher awareness that we have not yet experienced.

In bringing this book to a close, we will turn our attention to a very early relevant and prophetic comment by one of the first theoretical and quantum physicists of this century, Sir Arthur Eddington (1882–1944). Eddington made important contributions to the physics of stellar motion and evolution and was one of the first physicists to fully grasp the theory of relativity. Eddington had led the now famous expedition that photographed a solar eclipse in such a way that it offered the first proof of Einstein's relativity theory. As early as

1929, Eddington made the following comment, in support of a mystical-metaphysical inquiry into underlying reality:

> We have learned that the exploration of the external world by the methods of physical science leads not to a concrete reality but to a shadow world of symbols, beneath which those methods are unadapted for penetrating. Feeling that there must be more behind, we return to our starting point in human consciousness, the one center where more might become known. There (in immediate inward consciousness) we find other stirrings, other revelations than those conditioned by the world of symbols...Physics most strongly insists that its methods do not penetrate behind the symbols. Surely then, that mental and spiritual nature of ourselves, known in our minds by intimate contact transcending the methods of physics, supplies just that...which science is admittedly unable to give.

A thorough consideration of psychology, physics, and metaphysics demonstrates that there are qualities in the universe that unquestionably can be appreciated only by an awakened consciousness. This consciousness must become unified with the object it seeks: beauty, truth, and good. The experiential conviction that oneself and the universe are one, requires that we traverse meta-spaces much vaster than those spaces measured in the realm of modern physics, such as the spaces traversed by accelerated particles around their magnetic rings.

It has become quite clear that this kind of movement can be accomplished only in the meditative state, where separateness of self and object ceases to exist, and where illusions are eliminated because they are seen as devoid of any true cause (Tolle, 1999, 2001). Mysticism has had the synoptic vision that we have never left this order of unbounded wholeness, and that what we experience now as separation, disharmony, destruction, and death is but an unnecessary illusion. Ultimately, it has been mysticism's meditative state that has provided us with the vision that implicate consciousness and its timeless, unified, and perfectly peaceful order is our own true reality. True reality has no past; therefore, judgment held against the self and others is an illusion. This illusion serves to maintain observed reality, not the ultimate reality of perfect unity and harmony. Never before in man's history has he been so convincingly offered such an optimistic choice.

This choice is being simultaneously offered by those fields most regarded as arbiters of consciousness and meaning: psychology, quantum physics, and mysticism. While it is our observation that we are moving in the direction of a different, a truer reality, why not make a choice now, so we may move more quickly, individually, and collectively to this ultimate goal? This process can correctively displace that part of consciousness characterized by a conspiracy of collective misthought.

A *Course in Miracles* has made the following most relevant and insightful comment:

> You can either project limited consciousness or extend unlimited consciousness. It is up to you, but you must do one or the other, for that is the law of the mind, and you must look in before you look out. As you look within, you choose what you want for yourself. This process is reflected outward as observed reality, and you will accept it from the world, because you put it there by wanting it. When you think you are experiencing what you do not want, it is because you still want it and project this choice. This leads directly to disassociation, for it represents the acceptance of two incompatible goals, each perceived in a different place; separated from each because you made them different...You attack the real world every day and every hour and every minute, and yet you are surprised that you cannot see it. If you seek love in order to attack it, you will never find it. For if love is sharing, how can you find it except through itself? Offer it and it will come to you, because it is drawn to itself. But offer attack and love will remain hidden, for it can live only in peace. When you look in and decide to manifest only truth and love, you will see them both within and without.

Summary

With the foregoing in mind, we come to the inescapable conclusion that consciousness can be equated with meaning, and this meaning has temporarily taken the form of energy and matter. This equation can flow in either direction. It remains for each of us, and all of us, to decide in which direction it will proceed. We can choose to remain either in an unreality of material, entropy, and destruction, or alter our collective consciousness and awaken to choose a truer reality of timelessness, peace, and unity. Expressed another way, we can experience a reality of beauty, truth, and good.

History has shown that a dedicated intention to accomplish this will ultimately be successful.

Questions and Answers

Q: You have indicated that guilt and forgiveness are central issues in ascending the ladder of consciousness. Could you provide some guiding principles from your own point of view?

A: I believe that the ego arises out of opposition to the will of God. The central error is the choice to identify with the evanescent and delusional decision to replace God's authority as first creator with our authority as first creator. This fugitive thought has been given time, locality, and material, and it represents our choice as a separated self, or ego. Examination reveals that such a thought is illogical, unnecessary, and psychotic. It is impossible to separate from one's Source.

We share in all aspects of our Source, except in our own creation. In fact, by being second, we are really first. If we share all attributes of our Creator, why should we assign any value to creating ourselves? When the mind chooses the ego, thereby identifying with the body, it misappropriates the power shared with God by projecting the power into a variety of capabilities that imitate the mind's true function. The ego develops its own logic, with capabilities that serve the purpose of separation. The apparent use of reason—such as thinking, learning, imaging, and memory—are in fact only functions of a brain in a body. These functions serve the ego well by using the logic of its "reason" to defend and support the belief that our separation is real and life outside Heaven is not impossible. Guilt is generated in the ego's system by seeing its position as an attack on God. Guilt thus attains the status of a stable illusion. In order for one to move toward pure consciousness, guilt must be removed. For this to take place, it is necessary that we change our minds by the process of forgiveness. We are in fact forgiving something that did not and could not have happened (the impossible choice to be something other than part of God's Mind). This "attack on God" can only be illusory. In an attempt to hide and deny our inner guilt, this guilt has been displaced out onto the world, where we view it as alien to the self. We try to gain further distance from guilt by attacking others but succeed only in attacking ourselves, since there is really no separation. *A Course in Miracles* states:

> The ego's plan for your salvation centers on holding grievances. It maintains that if someone else spoke or acted differently; if some external event or circumstance were changed, we would be saved. Thus the source of salvation is constantly perceived as outside the self. Each grievance you hold is a declaration or an assertion in which you believe once again that if external circumstances were different you would be completely happy. The change of mind necessary for salvation is thus demanded of everyone and everything except yourself...The Holy Spirit is in both your minds, you and your brother and He are One, because there is no gap that separates His Oneness from itself. The gap between your bodies' matters not for what is joined in Him is always one.

It is totally irrelevant for the purpose of forgiveness that there be or not be a "gap between bodies," for we are already joined. Forgiveness functions to undo the apparent gap that exists between our minds, allowing us to return to the remembrance of our oneness with each other and our Source.

In conclusion, I've come to regard objectivity as an inadequate construction for achieving a vision of true reality. Logical processes are based entirely on belief and subjectivity. The inherent authority of truth in any concept can only have subjective value. What convinces one person is dismissed by another, for what is credible can only be an experience of subjective decision beyond definition. To know and realize God can only be a subjective experience. It remains a truth that *cannot* be concluded by reason. This Truth becomes knowable only by the *identity of being it*.

Glossary

Algorithm — A special step-by-step method of solving a kind of mathematical problem.

Affect — The affects are the most direct psychic derivatives of the instincts and are psychic representatives of the various bodily changes by means of which the drives manifest themselves. The affects regularly attach themselves to ideas and other psychic formations to which they did not originally belong, and as result their origin and meaning remain hidden from consciousness. If an affect is completely suppressed, it may appear not as an emotion but rather as physical changes or innovations, such as perspiration, tachycardia, paresthesia, etc.

Antithesis — Direct contrast or opposition.

Bell's theorem — The correlations discovered by physicist John Stewart Bell (1928–90) between simultaneous measurements of two widely separated particles on the assumption of the locality of hidden variables, in which a limit is set for information transfer (186,000 miles per second, the speed of light). If this limit is exceeded, reality is not governed by the rules of linearity and locality. Alain Aspect experimentally checked a pair of photons in 1983 and found the speed limit to be violated. Therefore, it is agreed that any model of reality must be nonlocal.

Big Bang theory — The theory that the universe was created by a gigantic explosion from a singularity about eighteen billion years ago. Its main confirmation is the detection of black-body background radiation left over from the original event. The theory states that there was no preexisting space or time before the original bang.

Bohm, David Joseph (1917–92) — One of the most eminent quantum physicists of the twentieth century. Bohm was one of the few physicists who early refuted the flawed proof of the mathematician John von Neumann, who held the position that hidden variables in physics did not exist. Bohm postulated the concepts of explicate, implicate, and superimplicate to describe connected but different orders of expanding consciousness. In 1952, Bohm did the impossible, constructing a model of the electron with innate attributes whose behavior matched the predictions of quantum theory, supporting the existence of hidden variables and thereby nullifying von Newmann's earlier proof. Bohm thought that the mind contained representations of the outside world and was a hologram within the greater holoverse.

Cartesian — A concept proposed by the French philosopher, mathematician, and scientist René Descartes (1596–1650), which is based on a reality of the physical world that is mechanistic, mathematical, and entirely divorced from the mind.

Character — The constellation of relatively fixed personality traits and attributes that govern a person's habitual modes of response.

Cognition — The mental process or faculty by which knowledge is acquired; that which comes to be known, as through perception, reasoning, or intuition; knowledge.

Cognitive — Refers to the mental processes of comprehension, judgment, memory, and reasoning, as opposed to emotional and volitional processes.

Collective unconscious or unitive unconscious — That aspect of our consciousness that transcends space, time, and culture, of which we are not aware. A concept first introduced by the psychoanalyst Carl Jung.

Complementarity principle — Theory proposed by the quantum physicist Niels Bohr (1885–1962), that particles in microscopic systems behave simultaneously as waves and as particles.

Consciousness — In this text, a particular quality of being, where there is an infinite spectrum of layers that are not separate but are mutually interpenetrated. Bohm refers to these layers as implicate orders; Wilber, as levels. All would agree that each individual layer is an aspect of an underlying whole and that all layers are accessible with the proper focus. All of us, conscious or not, are on a journey to widen our focus.

Copenhagen interpretation — The standard interpretation of quantum mechanics, developed by Bohr and Heisenberg. It is based on the concepts of the probability interpretation and the principles of uncertainty, complementarity, correspondence, and the inseparability of the quantum system and its measuring apparatus.

Countertransference — A process in psychoanalysis where the patient's personality or material he produces, comes to represent an object from the analyst's past, onto which past feelings and wishes are projected.

Delusional — A false belief out of keeping with the individual's level of knowledge and his cultural group. The belief results from unconscious needs and is maintained against logical argument and objective contradictory evidence.

Determinism — A doctrine that acts of the will, occurrences in nature, or social psychological phenomena are determined by antecedent causes.

Dialectic — A discussion and reasoning by dialogue as a method of intellectual investigation that involves systematic reasoning and usually seeks to resolve conflict by a process of eliciting the truth.

Disassociated — A process of splitting off some past or component of mental activity, which then acts as an independent unit of mental life.

Displacement — A defense mechanism, operating unconsciously, in which an emotion is transferred, or displaced, from its original object to a more acceptable substitute.

Dissipative — A theory of physics developed by physicist Ilya Prigogine, where systems deconstruct, or fall apart, with the passage of time in order to fall together or to integrate at a higher order at a later time, either on an observed or on an unobserved level.

Double-slit experiment — The classic experiment by Thomas Young in the early nineteenth century for determining characteristics of waves. A wave of light, for example, is split by passing it through two slits in a screen to make an interference pattern on a photographic plate or fluorescent screen. In the words of Richard Feynman, "It encapsulates the central mystery of quantum physics...it's a phenomenon which is...absolutely impossible, to explain in any classical way, and has in it...the basic peculiarities of all quantum mechanics."

Drive — A strong motivating tendency or instinct, especially of sexual or aggressive nature, that prompts activity toward a particular end.

Ego — In psychoanalytic theory, one of the three major divisions in the model of the psychic apparatus, the others being the id and superego. The ego represents the sum of certain mental mechanisms, such as perception, memory, and specific defense mechanisms. The ego serves to mediate between the demands of primitive instinctual drives (the id), of internalized parental and social prohibitions (the superego), and reality. The compromises between these forces achieved by the ego tend to resolve intrapsychic conflict and serve an adaptive and executive function.

Egocentric — Thinking or acting with the view that one's self is the center, object, and norm of all experience. Self-centered.

Einstein, Albert (1879–1955) — Perhaps the most famous physicist who ever lived, he is the discoverer of the relativity theories. He was a major contributor to quantum theory, including the basic ideas of wave-particle duality and probability. In his later years, he found the instrumentalist (and positivistic) trend of the interpretation of quantum physics distasteful to his scientific beliefs.

Electromagnetics — Having both electrical and magnetic properties.

Electrons — A particle of matter with a negative electric charge.

Entropy — In statistical mechanics, a measure of disorder in a closed system. This type of disorder refers to the evenness of the distribution of energy: An increase in evenness is an increase of disorder. According to the second law of thermodynamics, entropy never decreases in a total system. If the universe is an isolated system, it perpetually increases in disorder.

Empiricism — The doctrine that knowledge is obtainable only by direct experience through the physical senses.

Epigenetic — Development of an organism through a series of processes in which unorganized masses differentiate into specific systems.

Epiphenomenon — A phenomenon that arises from the organization of a particular system, but is not present in its constituent parts.

Epistemological — A philosophic theory of the method or basis of human knowledge.

EPR paradox — A paradox invented by Einstein, Podolsky, and Rosen to establish the incompleteness of quantum mechanics. Instead, the paradox paved the way for the experimental proof of quantum nonlocality.

EPR correlation — A phase relationship that persists even at a distance between two quantum objects that have interacted for a period and then have stopped interacting. The EPR correlation corresponds to the potential nonlocal influence between these objects.

Equation — A mathematical statement that two expressions are equal.

Explicate order — Physicist David Bohm's term for the domain referred to by Cartesian coordinates (locality in space-time). It displays the separateness and independence of fundamental constituents and is manifest or visible (directly or with instruments). It is secondary to the implicate order, which unfolds to create the explicate order and enfolds to give guidance to itself.

Extrapolation — To make an estimate of something unknown and outside the range of one's data (on the basis of available data).

Fermion — A particle or pattern of string vibrations with half a whole odd number amount of spin, typically a matter particle. Named after physicist Enrico Fermi, who in 1942 produced the first controlled nuclear chain reaction.

Feynman, Richard Phillips (1918–88) — Widely regarded as the greatest physicist of his generation, ranking alongside Isaac Newton and Albert Einstein. He reformulated quantum mechanics and placed it on a secure logical foundation in which classical mechanics is logically incorporated. He made numerous contributions in the areas of superfluidity and strong and weak forces, and to a quantum theory of gravitation. He also developed the most lucid and complete version of quantum electrodynamics (QED), which along with the general theory of relativity constitutes the two most successful and established theories in physics.

Field — In physics, a region throughout which a force may be exerted; examples are the gravitational, electric, and magnetic fields that surround, respectively, masses, electric charges, and magnets. Fields are used to describe all cases in which two bodies separated in space exert a force on each other.

Free association — In psychoanalytic therapy, spontaneous, uncensored verbalization by the patient of whatever comes to mind, which is intended to reveal aspects of the unconscious.

Heisenberg, Werner Karl (1901–76) — A German physicist and codiscoverer of quantum mechanics, and the author of the uncertainty principle. He was perhaps the only one among the founders of quantum physics to really understand and advocate the idealist nature of quantum metaphysics. His discovery of quantum mechanics is widely regarded as one of the most creative events in the history of physics.

Hidden variables — Unknown (hidden) parameters that are posited by Bohm and others to restore determinism to quantum mechanics; according to Bell's theorem, any hidden variables that reside in a world outside of space-time are inconsistent with material realism.

Holism — View that an organic or integrated whole has an independent reality, which cannot be understood simply through an understanding of its parts.

Holistic — In psychiatry, an approach to the study of the individual as a unique entity, rather than as an aggregate of physiological, psychological, and social characteristics.

Hologram — Three-dimensional photographic image reproduced from a pattern of interference produced by a split coherent beam of radiation (as of a laser).

Holomovement — In David Bohm's terminology, the total ground of that which is manifest. The manifest is embedded in the holomovement, which exhibits a basic movement of unfolding and enfolding.

Holoverse — Another term used by David Bohm that represents the holographic qualities, in which each increment of the universe contains all of the information necessary to construct the whole of the universe.

Humanism — A philosophy or attitude that addresses itself exclusively to human as opposed to divine or supernatural concerns, often coupled with the belief that man is capable of reaching self-fulfillment without divine aid.

Immanence — The quality of being inherent.

Implicate order — The basic order, according to Bohm, from which our three dimensional world springs. It is multidimensional, and its connections are independent of space and time. The implicate order is identified with the wave function in quantum theory.

Infrastructure — An underlying base or supporting structure.

Interference pattern — The pattern of reinforcement of a wave disturbance in some places, and cancellation in others, that is produced by the superposition of two (or more) waves.

Locality/nonlocality — The condition that defines the causal relationship between events. In a local reality, information cannot travel faster than light. In a nonlocal reality, objects can influence each other instantaneously. Bell's theorem indicates that if the world is made up of separate objects, they must have nonlocal connections. Bohm's concept of a quantum potential supports the concept that such connections exist.

Materialism — A theory that physical matter is the only reality and that all being and processes and phenomena can be explained as manifestations or results of matter.

Matter — According to David Bohm and Ken Wilber, matter is condensed consciousness. The unfolding and enfolding process created successive localized manifestations that appear to our senses and instruments as physical forms.

Metaphysics — The study of being (ontology), which often includes the study of the structure of the universe (cosmology). This branch of philosophy systematically investigates the nature of first principles and problems of ultimate reality.

Monistic idealism — The philosophy that defines consciousness as the primary reality, as the ground of all being. The objects of empirical reality are all epiphenomena of consciousness that arise from the modifications of consciousness. There is no self-nature in either the subject or the object of a conscious experience apart from consciousness.

Mystic — One who is capable of transcending the physical realm of space and time through a state of consciousness that extends beyond the restriction of the intellect to unity with the absolute.

Mysticism — A spiritual discipline aiming at union with the Divine through deep meditation or trancelike contemplation. Any belief in the existence of realities beyond perceptual or intellectual apprehension but central to being and directly accessible by intuition.

Narcissism — A term used to indicate self-love or an exaggerated overestimation of the self.

Neutron — Subatomic particle having no electric charge and a mass very close to that of a proton.

Newton, Isaac (1642–1727) — Now considered to be the founder of classical mechanics of particle physics with his three laws of mechanics or motion and his theory of gravity. The three laws of motion state that: a body continues at a state of rest or of uniform motion in a straight line unless it is acted upon by external forces; the rate of change of momentum of a body is proportional to the external force; any force (action) on a system gives rise to an equal and opposite force (reaction).

Newtonian physics — Sir Isaac Newton discovered the law of universal gravitation in 1666, which explained the motion of planets and falling bodies. He also discovered that white light is composed of every color in the spectrum and theorized that light is composed of particles.

Ontological — The branch of metaphysics dealing with the nature of existence.

Paradigm — In science, a conceptual framework endorsed by the large majority of the scientific community that presents problems to solve and defines boundaries within which solutions are sought. The term is often applied to the societal realm, in which it refers to the concepts and values of a community that shape its perception of reality.

Paradox — A seemingly contradictory statement that may nonetheless be true.

Participator — An observer who not only observes an occurrence but changes it by the act of observation, which in quantum reality represents observer/observed fusion.

Polarization — Producing or acquiring the condition of having magnetic poles (negative and positive).

Polarization correlation — Two photons related in phase so that if one's polarization is collapsed along a certain axis (as manifested by observation), the other's polarization is also collapsed along the same axis (as determined by observation) regardless of the distance between the photons.

Positron — The positively charged, antimatter counterpart to the electron.

Pythagorean — A philosophy of Pythagoras, chiefly distinguished by its description of reality in terms of arithmetical relationships and the doctrine of the transmigration of souls.

Quantum — Discrete unit of energy.

Quantum field theory — The theory that results from the application of quantum mechanics to the behavior of a field. In particle physics, all fields were treated as contiguous, extending to all points of space. However, when quantum mechanics was applied to the field it became quantized and exhibited definite energy states. This development allowed for a particle interpretation and made the wave-particle interpretation and wave-particle duality somewhat more understandable.

Quantum foam — John A. Wheeler's proposed picture of space as composed of microscopic bubbles forming what can be conceptualized as a carpet of foam, consisting of nonmaterial intelligence.

Quantum interconnectedness — Proposition that through the bubbles in the quantum foam all points in space and time are connected to all other points in space and time.

Quantum mechanical — Having to do with the workings or mechanics of atomic systems.

Quantum physics — Branch of physics that deals with the study of atomic systems.

Quantum potential — Connecting principle between quantum mechanical events that exists literally beyond space-time.

Quantum theory — A mathematical theory that describes the behavior of all systems (in principle) that is especially useful in conceptualizing atomic and subatomic realms. According to quantum theory, individual events are inherently unpredictable and inseparable and also are observer dependent in the view of many quantum physicists.

Quark — Any of a hypothetical set of fermions having electric charges of magnitude one-third or two-thirds that of the electron, proposed together with their antiparticles as the fundamental units of baryons and mesons.

Reality — All that is the case, including both local and nonlocal, immanent and transcendent. In contrast, the universe of space-time refers to the local, immanent aspect of reality.

Relativity theory — The theory of special relativity discovered by Einstein in 1905 that changed our concept of time from the Newtonian absolute time to a time existing in relation to motion. This theory explained the difference between accelerated and nonaccelerated systems and their relationship to gravity. As a consequence, space-time is seen not as rigid but as elastic, with its curvature determined by matter, energy, and gravity which then become a product of geometry rather than a force. These aspects of relativity theory resulted in the concepts of black holes, finite and unbounded universes, and even rips in space-time. The theory that postulates that the velocity of light is constant for all observers or sources. Therefore, physical laws are the same in all inertial frames of reference. From these assumptions, Einstein deduced the equivalence of mass and energy and the elasticity of space and time.

Repressing — A process by which memories, ideas, or fears are placed in the unconscious mind.

Repression — The active process of keeping out, ejecting and banishing from consciousness, ideas or impulses that are unacceptable to it.

Schrödinger, Erwin (1887–1961) — An Austrian physicist and codiscoverer with Heisenberg of quantum mechanics. He was opposed to the probability interpretation for quite some time. Later in life, he embraced some elements of the philosophy of monistic idealism.

Schrödinger's equation — An equation that describes the time evolution of a quantum state as represented by a quantum wave function. It is a complex differential equation that is completely deterministic for when a measurement is made; the quantum state is abruptly changed to one of a set of new possible states for which the probability of occurrence can be computed from the wave function.

Self-reference cosmology — View held by David Bohm that the universe exists ultimately as an unbroken whole in which all parts are simultaneously creating and being created by all other parts.

Sentient — Being capable of perceiving by the senses.

Superego — In psychoanalytic theory, that part of the personality associated with ethics, standards, and self-criticism. It is formed by the infant's identifications with important and esteemed persons in his early life, particularly parents. The supposed or actual wishes of these significant persons are taken over as part of the child's own personal standards to help form the conscience. In late life, they may become anachronistic and self-punitive, especially in neurotic persons.

Superimplicate order — A super-information field that serves as an orchestrating entity for the implicate order, determining its unfolding and enfolding process. This process then gives rise to the expressed material forms of the explicate order.

Superspace — The physicist John A. Wheeler's view is that the stage on which the space of the universe moves is certainly not space itself. The arena must be a larger object; which is not endowed with only three or four dimensions but with an infinite number of dimensions. He states that any single point in superspace represents an entire, three dimensional world. Space is regarded as composed of quantum foam and all matter in the universe is comprised of this one ultimate substance. The condition of material in superspace is super compaction.

Superunconscious—A plane of consciousness that is regarded as less accessible than the unconscious because it contains intensely fearful thoughts such as the fear of God's love or retaliation, completely delusional but strongly denied or repressed by minds that reside at the lowest levels of consciousness.

Theorem — A proposition to be proved by a chain of reasoning and analysis.

Transference — In psychoanalytic treatment, the projection of feelings, thoughts, and wishes onto the analyst, who has come to represent an object from the patient's past.

Uncertainty principle — The statement by Werner Heisenberg that it is impossible to measure a pair of noncompatible phenomena (e.g., position and momentum) both precisely and simultaneously, because all energy comes in quanta and all matter has both wave and particle aspects. An accurate measurement of position requires a short wavelength (high energy), and an accurate momentum measurement requires long wavelength (low energy).

Unified field theory — Any theory that describes all four forces and all of matter within a single, all-encompassing framework.

Wave function — Abstract line or function in configuration space representing the physical state of a system.

Wave packet — The configuration that results when a wave function is confined to a small region of space. This implies that the particle being described is confined to a fairly localized position.

Wave-particle duality — The twofold nature of the quantum objects, which can display wavelike and particle-like properties.

Wilber, Ken (1949–) — He is widely regarded as the most prominent biophilosopher of the twentieth century. His lucid studies integrate the areas of clinical and developmental psychologies, quantum physics, metaphysics, and expanding consciousness. He is seen as an important representative of transpersonal psychology, emerging from the humanistic psychology of the 1960s, which concerns itself explicitly with spirituality. In his work *Sex, Ecology, Spirituality: The Spirit of Evolution* (1995), he has criticized not only Western culture, but also counterculture movements such as the New Age movement. In his opinion, none of these approach the depth and detailed nature of the perennial philosophy; the conception of reality that lies at the heart of all major religions, which forms the background for all his writings. For the fundamental and pioneering nature of his insight, he as been called the Einstein of Consciousness.

Wholeness — Bohm's concept of an undivided, flowing movement, unbroken and all-encompassing but not complete and static. Division or analysis into an arrangement of objects and events can be appropriate for a limited description or display, but is not the primary reality or ground.

Zero-point energy — The irreducible quantity of energy, even at the minimum energy possible, of the discrete states of the quantum field—a result of the uncertainty principle. As a result of the uncertainty principle, a well-confined particle will begin to zero-point oscillate using the limits of space put on it to "create" energy. This means that the well-defined confinement of a particle in a cooled gas or liquid begins to make the gas or liquid boil even though the temperature is low. This "boiling point," which is given the special name of Einstein-Bose condensation temperature, was first discovered by Albert Einstein. This means that if the temperature is lowered still further past the condensation point, the gas will begin to behave in a bizarre manner in which the individual particles lose their separate identities. They cannot be said to be occupying such small spaces any longer. Instead, each particle fills the whole volume! For example, in this superfluid state the gas exhibits no resistance to the movement of an object through it. It has absolutely no viscosity. A similar situation occurs with certain supercooled metals called superconductors. These superconductors show remarkable electrical and magnetic properties. An electrical current, once started in a superconductor, continues without resistance. A magnetic field "caught" by the superconductor is totally enclosed by it. No magnetic-field lines leak out. This aspect of quantum physics has valuable applications to medical imaging, an area of this author's research and clinical application.

Bibliography

Anonymous, *A Course in Miracles,* Second Edition, The Foundation for Inner Peace, Mill Valley, California, 1996.

Aspect, A., A. Dalihard, and G. Rodger, *Physical Review Letters,* 39, December 1982.

Beck, A., *Cognitive Therapy and the Emotional Disorders,* International University Press, New York, 1976.

Beck, A., *Cognitive Therapy of Personality Disorders,* Guilford Press, New York, 1990.

Beck, A., A. Rush, and B. Shaw, *Cognitive Therapy of Depression,* Guilford Press, New York, 1979

Beynam, L., *The Emergent Paradigm in Science,* In Revision 1 and 2, Heldref Publications, Washington, D.C., 1978.

Bohm, D., "A New Theory of the Relationship of Mind and Matter," *Journal of the American Society for Psychological Research,* 80(2), 1986.

Bohm, D., "The Enfolding – Unfolding Universe," in *The Holographic Paradigm and Other Paradoxes,* K. Wilber, editor, Shambhala, Boston, 1982.

Bohm, D., "Hidden Variables and the Implicate Order," *Quantum Implications,* B. Hiley and H.D. Peat, editors, Routledge and Kegan, London, 1987.

Bohm, D., *Wholeness and the Implicate Order,* Routledge and Kegan, London, 1980.

Bohm, D., *On Dialogue,* David Bohm Seminars, PO. Box 1452, Ojai, California, 1990.

Bohm, D., and M. Edwards, *Changing Consciousness*, Harper, San Francisco, 1991.

Bohm, D., and B. Hiley, *The Undivided Universe*, Routledge, London, 1994.

Bohm, D., and F.D. Peat, *Science, Order and Creating*, Bantam Books, New York, 1987.

Bohr, N., *Atomic Physics and Human Knowledge*, John Wiley and Sons, New York, 1939.

Burns, D., *Feeling Good: The New Mood Therapy*, William Morrow and Co., New York, 1980

Cannon, Walter B., *The Wisdom of the Body*, W.W. Norton, New York, 1932.

Chopra, D., *Unconditional Life*, Bantam Books, New York, 1991.

Chopra, D., *Ageless Body, Timeless Mind*, Harmony Books, New York, 1993.

Conkling, M., "Sartre's Refutation of the Freudian Unconscious," *Review of Existential Psychology and Psychiatry*, 8: 86, 1968.

Davies, P., *The Mind of God*, Simon and Schuster, New York, 1992.

Davies, P., *Other Worlds*, Simon and Schuster, New York, 1980.

Davies, P., and J. Brown, *Ghost in the Atom*, Cambridge University Press, New York, 1992.

de Broglie, H., *Matter and Light*, Norton, New York, 1939.

DeWitt, B.S., "Quantum Mechanics and Reality," *Physics Today*, 23(9), 1970.

Dossey, L., *Recovering the Soul*, Bantam, New York, 1989.

Dossey, L., *Time, Space and Medicine*, Shambhala, Boston and London, 1985.

Dyson, F., "The Argument from Design," in *Disturbing the Universe*, Harper and Row, New York, 1979.

Eddington, A., *The Nature of the Physical World*, New York, Macmillan, 1929.

Einstein, A, quoted in Ruth Renya, *The Philosophy of Matter in The Atomic Era*, Asia Publishing House, New York, 1962.

Einstein, A., *Ideas and Opinions*, Carl Seeling, editor, Dell, New York, 1973.

Einstein, A., N. Rosen, and B. Podolsky, "Can Quantum Mechanical Description of Physical Reality Be Considered Complete?" *Physical Review*, 47, 1935.

Ellis, A., *A Guide for Rational Living in an Irrational World*, Prentice-Hall, Englewood Cliffs, New Jersey, 1971.

Fenichal, Otto, *The Psychoanalytic Theory of Neurosis*, W.W. Norton, New York, 1945.

Feynman, Richard P., *The Character of Physical Law*, MIT Press, Cambridge, 1967.

Feynman, Richard P., *The Feynman Lectures on Physics,* Volume 2, Addison-Wesley, Palo Alto, California, 1964.

Freud, S., *Beyond the Pleasure Principle,* W.W. Norton, New York, 1961.

Freud, S., *Standard Edition of the Complete Psychological Works of Sigmund Freud,* Hogarth Press, London, 1953–1956.

Friedman, N., *Bridging Science and Spirit,* Living Lakes, St. Louis, 1990.

Greenson, R., *The Technique and Practice of Psychoanalysis,* International University Press, New York, 1967.

Gribben, J., *Q is for Quantum,* An Encyclopedia of Particle Physics, Simon and Schuster, New York, 1998.

Grof, S., *Beyond the Brain,* State University of New York Press, Albany, New York, 1985.

Grotjohn, M., *Beyond Laughter,* International University Press, New York, 1958.

Hall, C., *A Primer of Freudian Psychology,* World Book Co., New York, 1954.

Hawkins, D., *The Eye of the I,* Veritas Press, Sedona, Arizona, 2001.

Hawkins, D., *I: Reality and Subjectivity,* Veritas Press, Sedona, Arizona, 2003.

Hawkins, D., *Power versus Force,* Hay House, Carlsbad, California, 1985.

Hayward, J., *Shifting Worlds, Changing Minds,* Shambhala, Boston, 1987.

Healy, W., A. Bronner, and M. Bowers, *The Structure and Meaning of Psychoanalysis,* Knopf, New York, 1930.

Heisenberg, W., *Physics and Philosophy,* Harper and Row Publishers, New York, 1944.

Herbert, N., *Quantum Reality,* Anchor/ Doubleday, Garden City, New York, 1985.

Holze, R., *Medicinische Physiologic de Scele,* 1852.

Huxley, A., *The Perennial Philosophy,* Harper and Row Publishers, New York, 1944.

Isaacs,W., *Dialogue: The Power of Collective Thinking,* Systems Thinker, April, 1993.

Jeans, James, *Autobiography of a Yogi,* cited in the book *Paramahansa Yogananda,* Self-Realization Fellowship, Los Angeles, 1983.

Kaufman, W., *Without Guilt and Justice,* Dell, New York, 1973.

Keepin,W., *David Bohm: A Life of Dialogue Between Science and Spirit,* Noetic Science Review, Summer,1994.

Kempf, E.J., *Psychopathology,* Mosby, St. Louis, Missouri, 1921.

Krishnamurti, J., and D. Bohm, *The Ending of Time,* Harper, San Francisco, 1985.

Lashley, Karl, *In Search of the Engram: Psychological Mechanisms in Animal Behavior*, Academic Press, New York, 1950.

Lazerowitz, M., "The Relevance of Psychoanalysis to Philosophy," in *Psychoanalysis Scientific Method and Philosophy*, S. Hooks, editor, Grove Press, New York, 1959.

Lorber, J., "Is Your Brain Really Necessary?" *Science*, 210, 1970.

Lukas, M., "The World According to Ilya Prigogine," *Quest*, 4: 10, 1980.

Margenau, Henry, *Miracle of Existence*, Shambhala, Boston, 1987.

Margenau, Henry, *Open Vistas*, Yale University Press, New Haven, 1961.

Margenau, Henry, "Fields in Physics and Biology," *Main Currents in Modern Thought*, 15-3, 1987.

Marmor, J., *Modern Psychoanalysis*, Basic Books, New York, 1968.

May, R., E. Angel, and H. Ellenberger, editors, *Existence*, Basic Books, New York, 1980.

Merleau, Ponty M., *The Phenomenology of Perception*, Routledge and Kagan, London, 1962.

Merleau, Ponty M., *The Structure of Human Behavior*, Beacon Press, Boston, 1963.

Mill, James, *Analysis of Phenomena of the Human Mind*,1829.

Muller, Robert, *New Genesis*, Doubleday, Garden City, New York, 1984.

Needleman, J., *Being in the World*, Harper Torch Books, New York, 1968.

Needleman, J., "Existential Psychoanalysis," in the *Encyclopedia of Philosophy*, P. Edwards, editor, Collier and Free Press, Glencoe, New York, 1967.

Newman, A., et al, *Why God Won't Go Away*, Balantine Books,New York 2001.

Ofman, W., "Existential Psychotherapy,"in *A Comprehensive Textbook of Psychiatry*, H. Kaplan,A. Freedman, and B. Sadock, Williams and Wilkins, Baltimore, 1980.

Ornstein, R., *The Experience of Time*, Penguin, New York, 1969.

Ouspensky, P.D., A New Model of the Universe: The Principles of the Psychological Method in its Application to Problems of Science, Religion, and Art, Dover, Mineola, New York, 1997.

Parrish, David, *Between Reality and Fantasy*, Grolnick and Barkin, editors, Aronson Press, New York, 1978.

Peat, F.D., *Synchronicity: The Bridge between Matter and Mind*, Bantam-Doubleday-Dell, New York, 1987.

Penrose, R., interviewed by David Freedman, "Quantum Consciousness," *Discovery*, June 1994

Penrose, R., "An Emperor Still Without Mind," in *Behavioral and Brain Sciences*, 16: 616-22, 1993.

Penrose, R., "A Search for the Missing Science of Consciousness," in *Shadows of the Mind,* Oxford University Press, New York, Melbourne, 1994.

Pribram, K., interviewed by Daniel Goleman, "Holographic Memory," *Psychology Today,* February 1979

Prigogine, I., and I. Stengers, *Order out of Chaos,* Bantam, New York, 1984.

Ring, K., *Life at Death,* Quill, New York, 1980.

Roheim, G., *The Original Function of Culture,* Doubleday, Garden City, New York, 1971.

Sartre, J.P., *Being and Nothingness,* Philosophical Library, New York, 1956.

Sartre, J.P., *Essays in Existentialism,* Citadel Press, New York, 1967.

Schrödinger, E., *Collected Papers on Wave Mechanisms,* Chelsea Publishing Co., New York, 1978.

Sheldrake, R., *A New Science of Life,* Second Edition, Park Street Press, Rochester, Vermont, 1995.

Shepherd, A.P., "Scientist of the Invisible,", quoted in *Future Science,* J. White and S. Krippner, Anchor, New York, 1977.

Smith, H., *Beyond the Post-Modern Mind,* Third Edition, Quest, Wheaton, Il, 2003.

Sperry, R.W., "Brain Bisection and Consciousness," in *Brain and Conscious Experience,* J.C. Eceles, editor, Springer, New York, 1966.

Stapp, H.D., "Correlation Experiments and the Proven Validity of Ordinary Ideas about the Physical World," *Physical Review,* D3, 1971.

Strupp, H.H., et al., Psychotherapy for Better or for Worse: The Problem of Negative Effects, Jason Aronson, New York, 1977.

Talbot, M., *The Holographic Universe,* Harper-Collins, New York, 1991.

Talbot, M., *Mysticism and the New Physics,* Arkana, Penguin, New York, 1993.

Tolle, E., *The Power of Now,* New World Library, Navato, California, 1999.

Tolle, E., *Practicing the Power of Now,* New World Library, Navato, California, 2001.

Trilling, L., "Authenticity and the Modern Unconscious," *Commentary,* 1971.

van den Berg, J., *A Different Existence,* Duquesne University Press, Pittsburgh, Pennsylvania, 1972.

Wagner, E.P., *Symmetries and Reflections,* Indiana University Press, Bloomington, 1967.

Wald, G., "Life and Mind in the Universe," *International Journal of Quantum Chemistry,* Volume II, Wiley and Sons, New York, 1984.

Wapnick, K., *A Vast Illusion: Time According to A Course in Miracles,* Foundation for a Course in Miracles Press, Roscoe, New York, 1990.

Weber, R., *Dialogues with Scientists and Sages: The Search for Unity*, Routledge and Kagan, London and New York, 1986.

Weber, R., "The Good, the True, the Beautiful: Are They Attributes of the Universe?" Vol. 32, No. 2, *Main Currents in Modern Thought*, 1975.

Weinberg, Steven, "Quarks," *Nova Series: Elementary Particle Physics*, Educational Television Network, November 23, 1974.

Wheeler, J., "The Black Hole and Beyond,", lecture delivered at Princeton University, November 7, 1974.

Wheeler, J., "The Universe as a Home for Man," in *Nature of Scientific Discovery*, O. Gingerich, Smithsonian Institution Press, Washington, D.C., 1975.

Wigner, E., quoted in *The Challenge of Chance*, Arthur Koestler, Random House, New York, 1974.

Wigner, E., *Symmetrics and Reflections*, Ox Bow Press, Woodbridge, Connecticut, 1979.

Wilber, K., *A Sociable God*, Shambhala, Boston, 1986.

Wilber, K., *The Atman Project: A Transpersonal View of Human Development*, Theosophical Publishing House, Wheaton, Illinois, 1980.

Wilber, K., *Eye to Eye: The Quest for the New Paradigm*, Anchor Press/Doubleday, Garden City, New York, 1983.

Wilber, K., *The Holographic Paradigm*, Shambhala, Boston, 1982.

Wilber, K., editor,_*Quantum Questions*, Shambhala, Boston, London, 1985.

Wundt, W.M., *Outlines of Psychological Psychology*, 1872.

Yankelovich, D., and W. Barrett, *Ego and Instincts*, Random House, New York, 1970.

Zimbardo, P., *The Cognitive Control of Motivation*, Scott Foresman, Chicago, 1969.

Zimmerman, J., "Time and Quantum Theory," in *The Voices of Time*, J.T. Fraser, George Braziller, New York, 1996.

Acknowledgements

I would like to express my gratitude to the many people who have contributed to this work; their kind support has made this book possible.

First of all, I am grateful for the love, encouragement, and wisdom of someone who modeled expanding consciousness throughout her life: my mother, Berdine Kotlarz. My wife, Pamela Gramer-Parrish, provided a rare and nourishing mixture of encouragement and tranquility, a most thoughtful and generous gift. Gary LaCroix, my dear friend, provided extraordinary support and assistance extending from the inception of the concept to the completion of the book. Jean Winter's knowledge of grammar, syntax, and composition was a helpful contribution. Sandra Parrish-Clinger rather unknowingly made a major contribution by way of her enduring belief in the basic goodness of everyone. My daughter Dr. Monique Parrish provided innumerable helpful psychological insights; my daughter Valerie Parrish's ongoing enthusiasm and affective warmth provided a positive backdrop for the identification of the developmental pathways traversed by expanding consciousness.

In the areas of consciousness and metaphysics, I would like to acknowledge the extended encouragement, wisdom, and guidance of Dr. Kenneth Wapnick. Without the previous creative and courageous work in medical psychology, quantum physics, and metaphysics of Sigmund Freud, David Bohm, and Ken Wilber, this book on consciousness could never have been written. Other major contributors were Hugh and Gayle Prather, Dr. Gerald Jampolsky, Dr. Wayne Dyer, Dr. David R. Hawkins, Dr. Jacob Teitelbaum, Dr. Robert Hedaya, Dr. Jeanette Murphy, Dr. Walter Conard, and Lynn Miller. Neva Klemme's typing as well as Dr. John Dungan's and Robert Spaeth's editing of the manuscript cannot be overestimated. Other important and valuable contributions were made by Ian Gordon, Cecelia and Larry Speer, Douglas Dickson, Dr. Daniel Eckstein, James Thomas and Anthony Bajoras, Stacey Powell, Connie Vitale, Jay Eldridge, and Susan Johnson.

About the Author

David Parrish, M. D., is a Freudian psychoanalyst who has researched developing consciousness in both early-child and adult development. He has studied quantum physics, psychology, and philosophy, and teaches in several of these areas. He began studying medicine in 1955, psychology and neurology in 1956, and psychosomatic medicine in 1958.

Parrish has an undergraduate degree in physics, and has done postgraduate studies both nationally and internationally in neurology and endocrinology. He also has had extensive hospice experience.

David Parrish's adult life has been shaped by both his professional commitment to science and medicine and his driving curiosity regarding metaphysics and expanding human consciousness. His research has resulted in dialogues with many of the philosophical, scientific, and spiritual thinkers of our time-among them, Richard Feynman, Kenneth Waprick, David Hawkins, and Ken Wilbur.

Dr. Parrish lives in Scottsdale, Arizona.

Sentient Publications, LLC publishes books on cultural creativity, experimental education, transformative spirituality, holistic health, new science, ecology, and other topics, approached from an integral viewpoint. Our authors are intensely interested in exploring the nature of life from fresh perspectives, addressing life's great questions, and fostering the full expression of the human potential. Sentient Publications' books arise from the spirit of inquiry and the richness of the inherent dialogue between writer and reader.

Our Culture Tools series is designed to give social catalyzers and cultural entrepreneurs the essential information, technology, and inspiration to forge a sustainable, creative, and compassionate world.

We are very interested in hearing from our readers. To direct suggestions or comments to us, or to be added to our mailing list, please contact:

SENTIENT PUBLICATIONS, LLC
1113 Spruce Street
Boulder, CO 80302
303-443-2188
contact@sentientpublications.com
www.sentientpublications.com